# VEHICLE LOCATION AND NAVIGATION SYSTEMS

# The Artech House ITS Series

For a complete listing of *The Artech House Mobile Communications Library,* turn to the back of this book.

# VEHICLE LOCATION AND NAVIGATION SYSTEMS

Yilin Zhao

Artech House, Inc.
Boston • London

Library of Congress Cataloging-in-Publication Data
Zhao, Yilin.
    Vehicle location and navigation systems /Yilin Zhao.
        p.    cm.
    Includes bibliographical references and index.
    ISBN 0-89006-861-5    (alk. paper)
        1. Intelligent Vehicle Highway Systems.    2. Motor vehicles—Automatic location
systems.    3. Electronics in navigation.
        I.  Title.
    TE228.3Z45    1997
    629.2'7—dc21                                                      97-4200
                                                                     CIP


British Library Cataloguing in Publication Data
Zhao, Yilin
    Vehicle location and navigation systems
    1.  Intelligent Vehicle Highway Systems  2.  Motor vehicles—Automatic location systems
    I. Title
    625.7'94


    ISBN 0-89006-861-5

**Cover design by Jennifer Makower**

International Standard Book Number: 0-89006-861-5
Library of Congress Catalog Card Number: 97-4200

10 9 8 7 6 5 4 3 2 1

*To everyone who provided help and support*

▼▼▼

# CONTENTS

*vii*

▼▼▼

# PREFACE

This book can be used as a reference for engineers, managers, or professionals who wish to familiarize themselves with or become professionally involved in vehicle location and navigation systems, or who have worked in a specific area of vehicle location and navigation and wish to better understand the entire field. The book can also be used as a textbook for undergraduates or beginning graduate students in the field of intelligent transportation systems (ITS, formerly IVHS) and related disciplines.

This is the first book to provide a detailed description of both the principles and practices of modern vehicle location and navigation systems in a single source. It combines information scattered among many different engineering fields into a single volume, a comprehensive introduction to the highly interdisciplinary field of ITS. It is based on several years of hands-on experience, and focuses on important aspects of each individual system module: its principles, implementation, integration, and combinatorial complexity. It permits a broad audience to learn about this fast-growing field without any prior knowledge. The concepts and algorithms are applicable to a broad range of location and navigation systems and form the foundation needed to understand, design, and implement advanced ITS. To help professionals with different backgrounds, the book attempts to limit the amount of mathematics used, and includes many newly developed technologies, practical examples, summary tables, and recent references. Even nonengineers should be able to understand much of the material.

The book is divided into two parts: Part I, Basic Modules (building blocks, Chapters 2 through 8), and Part II, Systems (integrated modules, Chapters 9 through 11). Chapter 1 is a general introduction to the vehicle location and navigation field, with a historical background and introduction to modern vehicle location and navigation systems. Chapters 2 through 8 describe the basic building modules from which systems are constructed. Chapter 2 discusses the digital map database generally used in vehicle location and navigation systems, as well as relevant reference coordinate systems and standards. Chapter 3 studies the positioning sensors and fusion methods used to locate the vehicle. Chapter 4 describes the map-matching algorithms employed for enhancing the positioning module and systems. Chapter 5 discusses various route-planning algorithms for finding the route that results in a minimum travel cost. Chapter 6 explains how to guide the vehicle along a given planned route that was derived on the basis of either the static travel cost or the dynamic travel cost. Chapter 7 discusses various human factors guidelines and technologies used for the human- machine interface, which interacts with the location and navigation computer and devices. Chapter 8 examines the different wireless communications technologies available for supporting vehicle communications.

Chapters 9 through 11 are devoted to systems. Chapter 9 discusses how to integrate the modules discussed in Part I to construct a variety of realistic autonomous vehicle location and navigation systems capable of operating without the support of a centralized remote computing and communications facility. Chapter 10 explains how a wireless communications module might be integrated with the vehicle and infrastructure to enhance in-vehicle systems using a centralized host computer. Chapter 11 uses an actual dynamic route guidance based system as a case study to shed some light on system architecture and implementation issues. The book concludes with Chapter 12, which describes past lessons learned and future directions projected.

Many useful summary tables and recent references are given, and many new technologies are introduced and practical questions discussed. Even though some technologies may not be discussed in detail here, the author has attempted to give comprehensive listings of available technologies wherever possible, in order to present a broad view of the field. When applicable, recent references are cited to direct readers to the latest information on and trends in the relevant technologies. Many newly developed technologies and practical questions are discussed in the book, including new technologies such as fuzzy-logic-based map-matching algorithms, the field-emitter display (FED), a newly invented software technology for map display that simulates a three-dimensional view, along with practical questions such as why the output of a GPS receiver might not match a certain digital map and why a certain navigation system might not work as expected right after an automatic car wash, and so forth. In addition, questions are provided at the end of each chapter to stimulate readers to think about the material learned in that chapter and to ponder these interesting problems themselves.

Because of recent widespread and rapid development of the Internet's World Wide Web (WWW) technology, the book also lists many addresses of ITS-related home pages in its references. The advantage of these "soft" pages is that you can read these references instantaneously if you have access to the Internet and a WWW browser. However, unlike "hard" printed papers and books, they may be unreachable due to computer or network problems. In addition, the URL addresses listed may be changed at any time by their owners, making them inaccessible from the listed addresses.

Due to the broad and rapidly expanding nature of vehicle location and navigation systems, it is difficult to cover the entire field in great detail. To limit the scope and conserve space, more emphasis has been given to the vehicle side of the system. The author intended the materials selected in the book to represent the most fundamental and significant aspects of recent developments. However, relevant concepts and references may have been omitted. The author would appreciate your comments and seeks your advice for improvements to the book. He will also maintain a WWW page for online receipt of suggestions and on-line posting of any errors not detected and corrected before the book went to print. The URL address of this page is http://birch.dlut.edu.cn/~yzhao/.

▼ ▼ ▼

# ACKNOWLEDGMENTS

I would like first to thank Motorola, Inc., for their permission to publish this book. However, the opinions expressed in the book are the personal views of the author, and do not necessarily represent positions held by Motorola.

This book would not have been possible without the help and support of my family, friends, and colleagues. I would like to express my gratitude to all those who have contributed in one way or another to the completion of this book. In particular, I would like to thank the following people for their contributions: Allan M. Kirson, who carefully reviewed eleven chapters and made many useful suggestions in addition to his support; Steve Albrecht, who patiently reviewed six preliminary chapters; Phil Pollock and Larry Willis, who made valuable contributions to help meet the publication deadline; Paul D. Makinen, who copyedited the book with great enthusiasm; an anonymous reviewer who provided many constructive suggestions; and many other people who reviewed parts of the book in their respective specialties and are listed alphabetically: Michael Barnea, J. Blake Bullock, Yanming Cheng, John Dillenburg, Thomas E. Hayosh, Garry C. Hess, Kenneth B. Hohl, Wei-Wen Kao, Howard L. Kennedy, Daniel Schwartz, Witold Wojciechowski, and the ADVANCE Project Office.

I would also like to acknowledge the help given by Kan Chen, Paul Green, Jay Jayapalan, Cynthia Keesan, Huangsheng Li, Charles J. Mages, Don Mills, Jim Nickel, Vijay Ramanathan, Heinz Sodeikat, Chengyi Xu, and librarians Constance M. Bytnar and Mary Lou Kutscha, as well as the support provided by Robert H. Reuss, Mike Smith, Cindi Moreland, Robert P. Denaro, Paul L. Dowell, Steven F. Gillig, and Lawrence E. Connell. I apologize in case anyone's name was accidentally omitted. This book is dedicated to all of those who provided assistance and support.

In addition, I would like to thank my teachers both in China and America for my early training and their later influence and inspiration. Thanks also go to many other Motorola IVHS team members who helped me during various projects that we worked on together, Motorola employees who provided and verified product information, and the outside companies which supplied their products and technical information for the book.

Finally, special thanks go to my wife, Ming Xiang, for her assistance and her sacrifice in allowing me to spend thousands of hours of family time on the book, and to my son, Lianhan Zhao, for his understanding. My late parents, Yueduan Lin and Weiduo Zhao (spelled Wei-to Chao in 1948, when he graduated from Case Western Reserve University in Cleveland), gave me all their love and were a great source of inspiration. Their encouragement, guidance and support will always be with me.

# CHAPTER 1
▼▼▼

# INTRODUCTION

Road transportation systems have undergone considerable increases in complexity and congestion. In particular, surface vehicle ownership and the use of vehicles are growing at rates much higher than the rate at which roads and other infrastructures are being expanded. One recent study found that, in the major urban areas of the United States alone, the total cost of traffic congestion exceeds $47.5 billion per year [1]. Traffic congestion wastes more than 14.35 billion liters (3.79 billion gallons) of fuel and 2.7 billion hours of work time per year. These numbers have continued to increase by 5% to 10% per year through the 1990s. Transportation authorities are increasingly turning to existing and new technologies to preserve mobility, improve road safety, and minimize congestion, pollution, and environmental impact. The roads and vehicles of tomorrow will see major changes brought about by the integration of computers, information, and communications technologies. This exciting new field is now known under the name intelligent transportation systems (ITS) (formerly known as intelligent vehicle highway systems (IVHS)).

The goal of ITS is to apply advanced technologies to make transportation operate more safely and efficiently, with less congestion, pollution, and environmental impact. The theory and practice of ITS are currently among the most intensely studied and promising areas in transportation, computer, information, communication, and systems science and engineering, and one that will certainly play a primary role in our future lives. If ITS lives up to its hopes, it will eventually affect virtually everyone in the world. In working toward this goal, ITS can take many different forms. In this book, we concentrate only on a specific portion of ITS: land-based vehicle location and navigation systems. However, the principles, concepts, and algorithms introduced in the book are applicable to a wide range of location and navigation systems, and form the foundation necessary to understand, design, and implement advanced ITS.

## 1.1  BRIEF HISTORY

Research on and development of vehicle location and navigation systems have a long history in our civilization, even though systems have only recently started to reach the world market. The earliest vehicle navigation system dates back thousands of years to the time when a south-pointing carriage was first invented in ancient China, around 2600 B.C. according to legend. The invention of this carriage has been written about since the San Guo (Three Kingdoms) period (220–280). During the Song Dynasty (960–1279), because of the invention of movable-type printing, detailed descriptions of this carriage were recorded and have been presented. Reconstructions were built in the San Guo period, 1027 and 1107. This carriage had a two-wheeled cart on which a human figure was mounted. This figure was kept continually pointed toward the south, no matter which way the cart was moving, through the action of a gear train. The basic mechanism of this carriage is similar to that of the modern differential odometer. The late Professor Zhenduo Wang of Beijing Normal University reconstructed this carriage from descriptions in a Yuan Dynasty book [2], as shown in Figure 1.1. For English literature and the working mechanism, refer to [3].

Almost in parallel with the invention of the south-pointing carriage, another interesting invention was the li-recording (distance-measuring) drum carriage. Like the south-pointing carriage, it had a set of gears. As the carriage moved, the gears turned to drive the arms of two wooden human figures. These figures sat on the carriage, one facing the drum and the other facing a gong. During the journey, one figure would strike the drum once every li, and another figure would strike the gong once every 10 li. A li is almost half of a kilometer. The fundamental mechanism of this carriage is similar to that of the modern odometer. From descriptions in the book mentioned above [2], Professor Wang reproduced this carriage, as shown in Figure 1.2.

As discussed in [3,4] and references therein, basic location and navigation technologies such as the odometer, differential odometer, and magnetic compass were invented about 2000 years ago. For the past century, these and other technologies have been gradually incorporated into modern automobiles. To learn more about the world's roads, bridges, and the vehicles that traverse them, refer to [5]. In the United States, the first automotive road map was published in 1895, and road signs were installed and roadways numbered early in this century. Mechanical route

**Figure 1.1** South-pointing carriage. (Courtesy of the Museum of Chinese History, Beijing.)

guidance devices, introduced around 1910, incorporated route map information in various forms such as sequential instructions printed on a turntable, punched in a rotating disk, or printed on a moving tape [6]. Each of these devices was driven by an odometer shaft in synchronization with the distance traveled along a route. Many mechanical vehicle route guidance devices patented from 1910 to 1920 provided explicit real-time route instructions automatically.

As roads became better marked and road maps became more accurate, interest in route guidance devices waned. Only a few developments occurred from the 1920s to the 1960s. During World War II, an electronic vehicle navigation system was developed in the United Sates for jeeps and other military vehicles [7]. This system had a magnetic compass whose needle position was read by a photocell. A servomechanism was driven by the compass output to rotate a mechanical shaft corresponding to the vehicle heading. This shaft was coupled to a mechanical computer that derived travel distance from an odometer and converted it into $x$ and $y$ components. Also driven by the compass shaft, these location components were used to automatically plot the vehicle trajectory on a map of appropriate scale.

In the late 1960s, an Electronic Route Guidance System (ERGS) was proposed by the U.S. Bureau of Public Roads (now the Federal Highway Administration) [8].

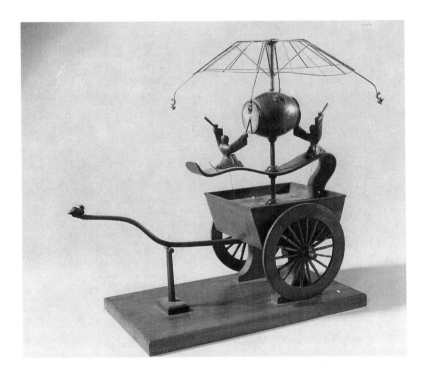

**Figure 1.2** Li-recording drum carriage. (Courtesy of the Museum of Chinese History, Beijing.)

The ERGS was a navigation system with wireless route guidance capability to control and distribute the traffic flow. A network of short-range beacons (proximity beacons) was used for two-way communications. An in-vehicle console with thumbwheel switches would permit the driver to enter a destination code. When approaching key intersections, the code would be transmitted from an on-vehicle transceiver to the beacon over a communication loop antenna embedded in the road surface of each approaching lane at the intersection and connected to a roadside controller via a coaxial cable. This controller was also connected to a centralized host for accessing traffic data. After receiving the destination code, the controller decoded it and planned a best route. The route guidance instructions were sent to the vehicle over the beacon before the vehicle left the vicinity of the loop antenna. These guidance instructions were then displayed on the vehicle's head-up display (HUD). Due to limited funding, this project was never fully implemented although successful tests were conducted. Despite never having been implemented, this project introduced the concept of centralized dynamic navigation (route guidance). Similar projects were later developed and tested in Japan and Germany during the 1970s. Note that much of the location and navigation terminology used in this section will be explained briefly in the following section and comprehensively in the following chapters.

In the early 1970s, an autonomous navigation system was introduced in the U.S. [9]. This system used a dead-reckoned positioning module assisted by a map-matching algorithm to locate the vehicle. A second version of the system could display route guidance instructions on a plasma display panel (PDP) once the vehicle's location along the route was confirmed. Similar systems were also developed by other groups in the United States and the United Kingdom. During the past 20 years, many key components and technologies for vehicle location and navigation have begun to mature.

During the past decade, vehicle location and navigation systems and other ITS-related systems have been rapidly gaining momentum worldwide. For detailed information on the projects and systems mentioned below, refer to [6,10–13], the numerous ITS-related World Wide Web (WWW) sites mentioned below, various ITS magazines and journals, as well as the proceedings of conferences organized by IEEE, the European Road Transport Telematics Implementation Coordination Organization (ERTICO), Vehicle, Road and Traffic Intelligence Society, Japan (VERTIS), and the Intelligent Transportation Society of America (ITS America).

In Japan, the ITS movement began in 1973 with the CACS project. The CACS project was similar in concept to the ERGS project. In the 1980s, an autonomous navigation system incorporating a color display monitor with a CD-ROM as the digital map storage device appeared on the market. Since then, a variety of navigation products have evolved, which include many new features such as map matching, use of a global positioning system (GPS) receiver, and voice guidance. More than 1.2 million vehicle navigation units have been sold to consumers. ITS field tests range from RACS and AMTICS in the 1980s to ATIS, VICS, and UTMS today. RACS was developed by the Public Research Institute of the Ministry of Construction and 25 private companies. AMTICS was initiated by the Traffic Management Technology Association in collaboration with private companies under the supervision of the National Police Agency (NPA). Later in 1991, RACS was combined with AMTICS to become the VICS program, with the support of an additional government agency, the Ministry of Posts and Telecommunications. VICS has been conducting tests and began deployment of technology in April 1996 for guiding vehicles to individual destinations, given real-time traffic information, using infrared beacons, microwave beacons, or FM broadcast subcarriers as transmission media (dynamic navigation). ATIS is being promoted by the Metropolitan Police Department, and uses cellular radio for vehicles and telephone lines for personal computers in fleet management centers. UTMS, initiated by NPA, plans to enhance existing traffic control systems by including traffic data collection and traffic information as well as traffic signal control and dynamic navigation.

In Europe, the ITS movement started in the late 1970s with the ALI project. The system was similar to the approach proposed in the ERGS project and tested in the CACS project. Autonomous navigation systems developed in the 1980s included CARIN and EVA. CARIN used dead reckoning with map matching and a color monitor for map display. It was the first navigation system to use a CD-ROM for map storage. EVA was demonstrated in 1983. In addition to dead reckoning, map matching, and turn-by-turn route guidance, it provided guidance to the driver in the form of both visual displays and synthesized voice output. ITS field tests (programs) known as PROMETHEUS and DRIVE have been operating since the mid 1980s. The former program was initiated by the automotive industry and has concentrated primarily on developments within the vehicle and its immediate environment. The latter program was initiated by the European Community and has been concerned mainly with infrastructure requirements. The two programs have been working together very closely and have many formal and informal links. Recently, PROMETHEUS has evolved into PROMOTE, and DRIVE has become a program under the name Telematics Applications for Transport and Environment.

In the United States, after the navigation systems proposed and developed in the late 1960s and early 1970s, an autonomous navigation system called Navigator was introduced commercially in the mid-1980s. It used a digital map database, dead reckoning with map matching, and a map display with icons showing current location and destination. Another autonomous navigation system named Guidestar, which incorporated a GPS receiver, appeared on the market in 1994. Many ITS field tests were conducted in the early 1990s, such as Pathfinder, TravTek, and ADVANCE. More projects are undergoing operational tests (see Table 10.1 in [10]) and even more are under construction. Most of these projects have been jointly proposed and developed by the public and private sectors. The purpose of these programs is to perform an initial assessment of the feasibility and utility of in-vehicle navigation, dynamic navigation, travel information, and various communication media, as well as other ITS concepts.

## 1.2 MODERN VEHICLE LOCATION AND NAVIGATION

As seen from the historical developments discussed earlier, human beings have never stopped pursuing the dream of great mobility. Thousands of years of civilization have led to the modern vehicle location and navigation systems of today. There are, and always will be, a range of systems from the very low end to the very high end. The low end might be very simple systems that detect the location of a vehicle or mobile device with human intervention, whereas the high end might be complex systems that navigate the vehicle automatically through the road network, assisted by real-time traffic information provided over a wireless communications network.

This book uses a modular approach to cover a broad range of systems. As we will learn in later chapters, a modern location and navigation system often consists of some or all of the modules shown in Figure 1.3. In other words, sophisticated location and navigation systems require integration of numerous functions and technologies. Therefore, we first focus on important aspects of each individual

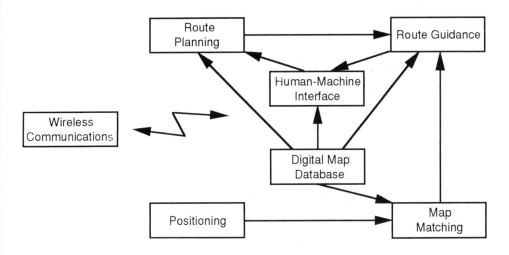

**Figure 1.3** Basic modules (building blocks) for a location and navigation system.

module, the principles underlying it, and the implementation of theses principles. We then examine how systems can be constructed from these modules, discuss the integration of these modules, and consider the additional complexity that arises when these modules are combined.

The modules depicted in Figure 1.3 can take different forms. Some of the modules were briefly described in the preceding section. These modules can be implemented by different hardware and software components. We now summarize the function of each individual module.

A digital map database contains digitized map information with a predefined format, which can be processed by a computer for map-related functions such as identifying and giving locations, road classifications, traffic regulations, and travel information. Because a map is used to represent the geometry of the Earth's surface, we must understand the relevant reference coordinate systems used by different map databases in order to develop map-related functions properly. To ease the pain of dealing with different digital maps, standards have been developed to guide development and applications. A positioning module fuses different sensor outputs or uses radio signals for automatic determination of the position of the vehicle or a mobile device to identify the road traveled and each intersection approached. A typical stand-alone technique is dead reckoning, and a typical radio-signal-based technique uses GPS receivers. A variety of fusion methods have been developed for integrated sensors. *Map matching* is a method of matching the position (or trajectory) measured or received by a positioning module to a position associated with a location (or route) on a map provided by the map database module. This technique can improve the accuracy of the positioning module, provided the map database is of reasonable precision. A typical precision requirement for urban areas is within 15m of "ground truth."

_Route planning_ is the process of helping vehicle drivers plan a route prior to or during their journey, based on a given map provided by the map database module, if available, along with real-time traffic information received via a wireless communications network. A commonly used technique is to find a minimum-travel-cost route based on criteria such as time, distance, and complexity. _Route guidance_ is the process of guiding the driver along the route generated by the route-planning module. It requires the help of an accurate positioning and map database in order to determine current vehicle position and generate proper real-time guidance instructions, often turn by turn. A human-machine interface allows users to interact with the location and navigation computer and devices. Map displaying, route planning, route guidance, and other activities are directed to the machine and fed back to the user via this module. A wireless communications module further improves the performance and increases the functionality of the system. Through one or more of various communications networks (for which more than 12 different technologies are available), a vehicle and its occupants or transportation management systems can receive updated traffic information or traffic reports, which make the in-vehicle system or whole road network operate more efficiently and safely. Of the several communications technologies discussed in this book, CDPD has been selected by the U.S. Joint ITS Architecture team as the preferred method for providing traffic and route-guidance information to motorists. Furthermore, a centralized facility will even be able to locate and navigate vehicles through its road network. All of these modules are discussed in individual detail in Part I, Chapters 2 through 8.

We will then learn how to integrate the above modules to construct working location and navigation systems in Part II, Chapters 9 through 11. Many practical systems are discussed to help readers understand these chapters. As in Part I, a system can take different forms, each of which may include a variety of different modules. First, autonomous location and navigation systems, such as vehicle location (or tracking) systems and navigation systems with no remote host or centralized computing facilities, are discussed. One of the main characteristics of this type of system is that all location and navigation capabilities are placed solely on the vehicle. Second, centralized location and navigation systems are examined. These systems utilize communications networks, host facilities, and other infrastructure to perform multivehicle location and navigation. For any specific system, three design issues must be considered: placement of location and navigation capabilities (on a host in a fixed infrastructure or on each individual vehicle), location accuracy and update frequency, and selection of wireless communications technologies. Third, a working dynamic route-guidance system is examined in detail. Unlike a static system, a dynamic system uses databases updated by real-time traffic information to guide the vehicle. This study will build on the knowledge and techniques learned in the previous chapters and demonstrate how they are applied in the real world. We will also discuss several system architecture, design, and implementation issues. Finally, in Chapter 12, we discuss lessons learned from past experience and predict future directions for vehicle location and navigation systems.

As stated, many modern systems consist of some or all of the modules in Figure 1.3. For instance, an autonomous in-vehicle navigation system may exclude only the wireless communications module. It may interact with a map database to plan a minimum-travel-cost route based on a given destination specified by the driver. Knowing the route, it may work with the positioning module and the map-matching module to guide the driver to follow a particular route to the destination. The guidance instructions may consist of voice, audible tones, and/or display text and graphics. Integration with a wireless communications network and a centralized host, which takes over some of the location or navigation functions, produces a centralized navigation system.

So that readers might better understand the various modules and systems, some current or planned products or systems are mentioned. Such information was gathered via extensive literature searches, direct phone interviews, or proofreadings by manufacturers. This information is believed correct at the time of publication. However, it may be changed or updated anytime during any phase of product or system development. In addition, to concentrate on the main theme of the book, we emphasize the modules used in the vehicle and systems employed in the vehicle. Application to other system architectures should not pose too much of a challenge once the reader has mastered the fundamental principles discussed here.

Despite our modular approach, readers must keep in mind that a top-down approach is generally used in product development. A top-down approach typically includes identifying system requirements, determining functions and system architecture, specifying appropriate modules, selecting hardware components and software tools, and designing, implementing, integrating, and testing the system. To facilitate our study, we will introduce the various individual location and navigation modules first, but this does not mean we can ignore issues of system architecture and integration. For any location and navigation system to be successful, it must be based on a good architecture. This system architecture must provide a stable basis on which the system can further evolve, through adding or upgrading individual modules in response to changing requirements and operating environments over the normal lifetime of the system. Sophisticated systems, in particular, are expected to have complex system integration. Without proper processes and management, system integration can be delayed considerably, even when the individual modules function properly.

Because of explosive technological advances, modern vehicle location and navigation systems have improved greatly during the last decade. A variety of communication tools has been used to educate the public and to promote further technical advances. Lately, dozens of Internet WWW pages have emerged to share information in this exciting new field. Interested readers can browse sites worldwide for instant research and development project status, the latest news, transportation statistics, real-time road traffic conditions, and much more. A few sample sites are listed at the end of this chapter [14–26].

Market studies have predicted great potential for vehicle location and navigation equipment and systems. One study estimates that by the year 2000 the world market for vehicle navigation units integrated with GPS receivers will be $3 billion (Fig. 1.4). Similar studies have been conducted in their respective regions by the Japan GPS Council, Nikko Research Center, Elsevier Advanced Technology, Allied Business Intelligence, Frost & Sullivan, NevTech, SRI, the University of Michigan, and many other organizations. One Japanese forecast predicts a total sales of 3 million units [27]; one European forecast estimates a total sales of 2.5 million units [28]; and one American forecast projects a total sales of 22.5 million units [29] for in-vehicle navigation systems in the year 2005 alone. Assuming that the unit cost will be $500 by that time, this leads to a total sales volume of $1.5 billion, $1.25 billion, and $11.25 billion, respectively, in each of these regions.

Vehicle location and navigation systems have already gained great user acceptance worldwide. In Japan, sales of vehicle navigation systems to consumers reached 530,000 units in 1995, up 73.2% from 1994, as reported by the Japan Electronic Industries Association. The ERTICO study, based on field tests in Europe, has confirmed high levels of user acceptance for in-vehicle route guidance and RDS-TMC (Radio Data System-Traffic Message Channel) systems [30]. Consumers are

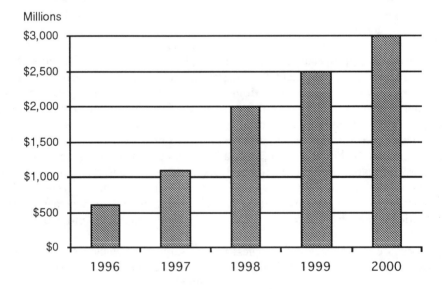

**Figure 1.4**  Worldwide market projections of GPS-based in-vehicle navigation systems. (*Source:* U.S. GPS Industry Council.)

willing to pay $950 to $1900 for the route-guidance unit and $133 for the RDS-TMC equipment. In the United States, consumers are also willing to use vehicle location and navigation systems, as indicated by a recent study conducted in five main American metropolitan areas with 10 focus groups in October 1995 [31]. The groups were composed of cellular phone subscribers over age 25 from households with annual incomes of more than $30,000. This study found that, as shown in Figure 1.5, interest was strongest in the so-called "mayday systems," which allow vehicle occupants to put themselves in instant wireless connection with a service center for emergency attention or assistance, while automatically reporting the vehicle location (see detailed discussions and the recently debuted products in Section 10.4).

All these market studies indicate that the vehicle location and navigation field has a promising future and should be economically rewarding. With this historical background and basic information on vehicle location and navigation systems in mind, we now proceed to a detailed description of these systems with the aim of

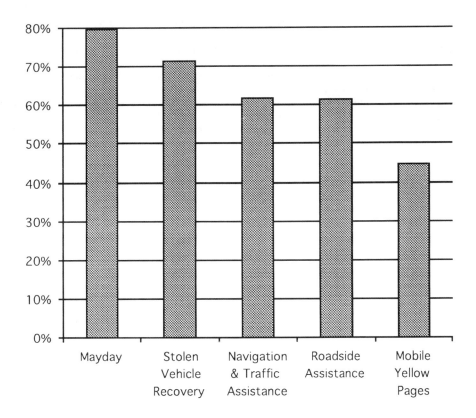

**Figure 1.5** American consumer interest in location-based mobile services. (*Source:* [4].)

providing the reader with a better understanding of the underlying principles and practices.

## References

[1] Texas Transportation Institute, *Urban Roadway Congestion 1982-1992, Volume 1: Annual Report*, College Station, TX: Texas Transportation Institute, Sep. 1995.

[2] T. Tuo and T. Alu (Eds.), *Song Shi* [History of The Song Dynasty, in Chinese], Yuan Dynasty, China, 1345.

[3] J. Needham, *Science and Civilization in China*, Vol. 4, Part II, Cambridge, UK: Cambridge University Press, 1965.

[4] J. Needham, *Science and Civilization in China*, Vol. 4, Part I, Cambridge, UK: Cambridge University Press, 1962.

[5] M. G. Lay, *Ways of the World: A History of the World's Roads and of the Vehicles That Used Them*, New Brunswick, NJ: Rutgers University Press, 1992.

[6] R. L. French, "From Chinese Chariots to Smart Cars: 2,000 Years of Vehicular Navigation," *Navigation: Journal of the Institute of Navigation*, Vol. 42, No. 1, Spring 1995, pp. 235–257.

[7] D. J. Faustman, "Automatic Map Tracer for Land Navigation," *Electronics*, Vol. 17, No. 11, Nov. 1944, pp. 94–99.

[8] D. A. Rosen, F. J. Mammano, and R. Favout, "An Electronic Route Guidance System for Highway Vehicles," *IEEE Trans. Vehicular Technology*, Feb. 1970, pp. 143–152.

[9] R. L. French and G. M. Lang, "Automatic Route Control System," *IEEE Trans. Vehicular Technology*, May 1973, pp. 36–41.

[10] I. Catling (Ed.), *Advanced Technology for Road Transport: IVHS and ATT*, Norwood, MA: Artech House, 1994.

[11] A. M. Parkes and S. Franzen (Eds.), *Driving Future Vehicles*, Washington, DC: Taylor & Francis, 1993.

[12] J. Walker (Ed.), *Mobile Information Systems*, Norwood, MA: Artech House, 1990.

[13] R. Whelan, *Smart Highways, Smart Cars*, Norwood, MA: Artech House, 1995.

[14] Bureau of Transportation Statistics, *Transportation Statistics*, http://www.bts.gov/, 1996.

[15] P. David, *Subway Navigator*, http://metro.jussieu.fr:10001/bin/cities/english, 1996.

[16] Georgia State DOT, *ATMS Home Page*, http://www.dot.state.ga.us/homeoffs/t_ops.www/, 1996.

[17] Institute of Transportation Engineers, *ITE on the Web*, http://www.ite.org/, 1996.

[18] Intelligent Transportation Society of America, *Access ITS America*, http://www.itsa.org/, 1996.

[19] Intelligent Transportation Systems Program, Princeton University, *Transportation Resources*, http://dragon.princeton.edu/~dhb/, 1996.

[20] Sytadin, *Real-Time Road Traffic Conditions* [in French], http://www.club-internet.fr:80/sytadin/map.html/, 1996.

[21] The United States Department of Transportation, *DOT Home Page*, http://www.dot.gov/, 1996.

[22] University of Leeds, *Institute for Transport Studies*, http://its02.leeds.ac.uk/, 1996.

[23] University of Valencia, *Spanish National Road Network*, http://carpanta.eleinf.uv.es/dgt/ap_autoI.html, 1996.

[24] Victorian State Government, *VicRoads Home Page*, http://www.vicnet.net.au/~vicroads/, 1996.

[25] Washington State DOT, *WSDOT Home Page*, http://www.wsdot.wa.gov/, 1996.

[26] J. Werner, *ITS Online*, http://www.itsonline.com/, 1996.

[27] T. Tachibama, *Car Navigation Forecast Data* [in Japanese], Autoparts Group, Nikko Research Center, Nov. 1994.

[28] T. Long, *Advanced Transport Telematics (ATT): A European Forecast Study for the In-Car System Market to 2005*, Oxford, UK: Elsevier Advanced Technology, 1996.

[29] P. Samuel, "Unit Prices Key to US Market for In-Vehicle Guidance," *ITS: intelligent transport systems,* June 1996, pp. 57–58.

[30] ERTICO, "ITS Studies Confirm User Acceptance," *ITS: intelligent transport systems*, Mar. 1996, pp. 41–42.

[31] C. J. Driscoll and D. Wolfe, *Location-Based Mobile Communications Services Marketing Research Study*, C. J. Driscoll & Associates and Wolfe & Company, May 1996.

# PART I

▼▼▼

# BASIC MODULES

Part I is devoted to a discussion of subsystems of vehicle location and navigation systems. As mentioned in Chapter 1, to make our study easier, we will treat these subsystems as separate modules. After studying these basic modules, we discuss how to integrate these modules into complete working systems in Part II. In other words, the intricacies of systems engineering will be dealt with later in the book.

In this part, we look at digital map databases in Chapter 2, positioning subsystems in Chapter 3, and map-matching algorithms in Chapter 4. We then discuss route-planning algorithms in Chapter 5, route-guidance methods in Chapter 6, human-machine interfaces in Chapter 7, and various existing communications technologies in Chapter 8.

# CHAPTER 2
▼▼▼

# DIGITAL MAP DATABASE MODULE

## 2.1 INTRODUCTION

A digital map database is an indispensable module for any vehicle location and navigation system that involves map-related functions. Without a map, it is very difficult for a traveler to explore an unfamiliar area and make correct decisions concerning the route. With a map as a medium, complex information can be communicated very easily.

A digital map database can provide many important functions for a vehicle location and navigation system. In short, it can help the system with the following tasks:

1. Display a map;
2. Locate an address or destination using a street address or nearby intersection;
3. Calculate a travel route;
4. Guide the driver along a precalculated route;
5. Match the sensor-detected vehicle trajectory with the known road network to determine more accurately the actual position of the vehicle;
6. Provide travel information such as travel guides, landmarks, hotel and restaurant information.

In general, there are two methods by which a computer can present a map to a traveler. One is to digitize a paper map (or photographs) using a scanner so it can be stored in and retrieved from computer memory as a digitized image or a raster-encoded structure (raster encoding). The other is to convert a paper map into a data structure or a vector-encoded structure so that it can be stored in and retrieved from computer memory easily, and presented in different ways to the user based on the functions or features required (vector encoding). Raster-encoded (scanned) maps can be easily produced and provide all the information contained in the original sources. They look exactly like the paper maps from which they are derived. On the other hand, vector-encoded maps require less storage and have faster access times. They are more flexible, easier to manipulate, and intrinsically relational in nature, since the data structure creates an implicit relationship between all of the various map elements. In the first (raster-encoded) method, values of each parameter of interest are provided for each cell (pixel) in a matrix over space. This type of map requires a large amount of storage and is difficult to use in many vehicle location and navigation applications that make use of mathematical models or calculations. Therefore, in this chapter we discuss only the second method, vector-encoded maps. Although we do not discuss the first method in detail, readers should be aware that this type of map still has applications despite its restrictions. For instance, it can be used for display purposes, for example, in vehicle location (tracking) systems where vehicle location or other information is superimposed on a map.

## 2.2  BASIC REPRESENTATIONS

Vector encoding is the representation of road network features using Cartesian geometry. A feature is denoted as an existing or planned item in the real world. Features are modeled by associating each feature with one or more primitives: points, lines, areas. A point is represented by two numbers with respect to a reference coordinate system. A line is represented by two connected points. An area is represented by a closed series of lines. The encoding is done in such a way that processing of the information may be automated using computers.

The purpose of the digital map database module is to define a network (Figure 2.1) of roads and associated attributes based on digital cartographic information, and then to compile it into a file or set of files that can be readily accessed by other modules in the location and navigation system. Digital cartographic information is geographic information that has been encoded in digital form. In other words, it is data digitized from existing maps or aerial photographs based on the primitives discussed earlier (and variations on these primitives). The attributes might include types of roads, street names, address ranges, expected driving speeds, connectedness, signs and signals, turn restrictions, points of interest, etc.

The digitized road network typically represents the road data using line segments whose endpoints (nodes) and shapes are defined in terms of latitude, longitude,

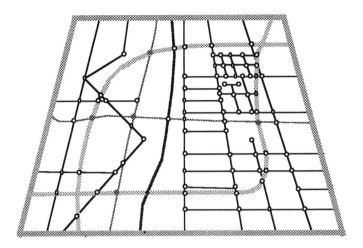

**Figure 2.1** A typical road network.

and sometimes relative altitude. To understand the digital map database better, we need to define nodes, segments, and shape points:

- A *node* is a cross point or an endpoint of a street/road and is used to represent an intersection or a dead-end of a road. It is commonly represented in map databases by a latitude and a longitude.
- A *segment* is a piece of roadway between two nodes and is used to represent fragments of roadways and other features. A node can be the endpoint of a segment, but not the endpoint of a shape segment (see the definitions of shape point and shape segment next).
- *Shape points* are ordered collections of points which map the curved portion of a given segment (excluding the endpoints) to a series of consecutive straight-line pieces (called *shape segments*) in such a way as to make the calculated segment distance close to the actual length.

Samples of segments, nodes, and shape points are shown in Figure 2.2. From these definitions, we now know that a road can typically be represented by a sequence of straight lines (vectors if travel directions are considered) selected to approximate the real curvature of the road.

A computer program, usually referred to as a *database compiler*, takes an input file containing raw digital cartographic information provided by a vendor and produces a file or a set of output files in a compact format suitable for rapid access by the location and navigation system (see Section 2.6 for a high-level description of this process). In addition to integrating and formatting segments and nodes over

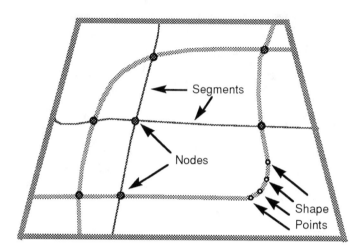

**Figure 2.2** Segments, nodes, and shape points in a map.

a preselected area, the compilation process can also involve attributes such as road names, road segment lengths, segment direction and speeds, address ranges for roads, and various other relevant additional data for use by the target navigation system. The final output data files must be in a format suitable for access by the run-time system on the target computer, the dedicated computer which manages all the activities of the location and navigation system.

A digital map database is a very important component of any location and navigation system. Once the position (coordinates) of a vehicle has been determined, the map provides all location-related features. The on-board digital map database can be used as a reference to visualize and locate vehicles. Map-matching algorithms can use the geometry and topology of road networks to help determine vehicle location (as discussed in Chapter 4). Therefore, digital map databases are more than aids for pictorial representation of location, they themselves can be a means of providing vehicle location.

## 2.3 REFERENCE COORDINATE SYSTEMS

A digital map represents the geometry of the Earth's surface along with the attributes (such as the coordinates) of various geographical features. A review of the terminology and coordinate systems used should be helpful, and will assist in explaining various map references used by different organizations. We will also explain why we cannot blindly mix maps with different references, how to transform between the various coordinate systems, and where to find good sources for further research.

First, we define some of the terms. The *geoid* is the surface along which the gravity potential is constant at all points and for which the gravity vector is normal to all points, that coincides with mean sea level extended through the continents (Figure 2.3). A *meridian* is a great circle on the surface of the Earth passing through the poles at the same longitude as the north-south reference line for the coordinate system. An *azimuth* is a horizontal angle measured through 360 degrees, from the north and reckoned clockwise from the meridian. An *ellipsoid* is a smooth closed surface whose plane sections are ellipses or circles. A *rotational ellipsoid* is the shape formed when an ellipse is rotated around one of its axes. The geometric parameters for an ellipsoid are as follows:

- Semi-major axis: $a$;
- Semi-minor axis: $b$;
- Flattening: $f = (a - b)/a$;
- Eccentricity: $e = (a^2 - b^2)^{1/2}/a$.

The Earth is an oblate rotational ellipsoid that is slightly flattened at the poles. There are many mathematical descriptions and coordinate systems for the Earth. To map a local geographical area accurately onto a map, many regional authorities will adopt a special regional coordinate system. The geodetic coordinates (latitudes, longitudes, and heights, described shortly) of these systems depend on the shape, size, and orientation of the reference ellipsoid. These parameters (along with others) serve as a reference or basis for the calculation of other quantities and, therefore, are called the *datum* of the coordinate system. It is desirable to have the rotational ellipsoid and the geoid match as closely as possible within the area under consideration. This explains why there are so many datums available. Some important ones are listed in Table 2.1. The full name of the datums given in Table 2.1 are listed below:

- WGS 84 stands for World Geodetic System 1984.
- NAD 83 stands for North American Datum 1983.

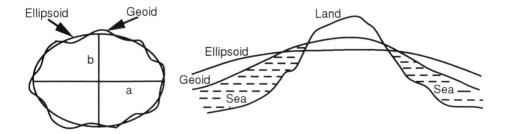

**Figure 2.3** Geoid and ellipsoid.

Table 2.1
Important Datums for the Earth

| Datum | Reference Ellipsoid | Ellipsoid Parameter | Datum Origin | Coverage |
|---|---|---|---|---|
| WGS 84 | WGS 84 | $a = 6378137m$ $f = 1/298.257223563$ | Geocenter (center of the Earth) | Global |
| WGS 72 | WGS 72 | $a = 6378135m$ $f = 1/298.26$ | Geocenter | Global |
| NAD 83 | GRS 80 | $a = 6378137m$ $f = 1/298.257222101$ | Geocenter | Regional |
| ED 50 | International Ellipsoid | $a = 6378388m$ $f = 1/297$ | Helmert's Tower, Potsdam, Germany | Regional |
| Pulkovo 42 | Krasovskiy 1940 | $a = 6378245m$ $f = 1/298.3$ | Center point of the Round Hall at Pulkovo Observatory, Russia | Regional |
| NAD 27 | Clarke 1866 | $a = 6378206.4m$ $f = 1/294.9786982$ | Meades Ranch, Kansas, USA | Regional |
| Tokyo Datum | Bessel 1841 | $a = 6377397.2m$ $f = 1/299.1527052$ | Center of the transit circle at the old Tokyo Observatory, Japan | Regional |

- GRS 80 stands for Geodetic Reference System 1980.
- WGS 72 stands for World Geodetic System 1972.
- ED 50 stands for European Datum 1950.
- NAD 27 stands for North American Datum 1927.

Note that the origin of the datum is a point fixed by definition and not necessarily the same as the origin of the coordinate system involved.

We will discuss two coordinate systems and one map projection: the geocentric coordinate system, the geodetic coordinate system, and the Universal Transverse Mercator (UTM) conformal map projection. Understanding these systems is important for incorporating map databases in this module as well as other modules to be discussed in other chapters. For instance, different map database vendors may supply databases defined by different reference ellipsoids or datums. The digital map database generally uses one reference ellipsoid as its coordinate reference while the positioning data received from a Global Positioning System (GPS, Section 3.4.2) receiver uses another reference ellipsoid (WGS 84) as its coordinate reference. Without proper conversions between these different coordinate systems, how can one expect the system to work as planned? Therefore, we must learn these different coordinate systems and know how to convert between them.

A geocentric coordinate system is a coordinate system with its origin at a specific point defined as the "center of the Earth," such as the center of mass (CM)

(Figure 2.4). It uses a polar (spherical) coordinate system in which the coordinates of a point are defined by the geocentric latitude $\phi$, the longitude $\lambda$, and the distance $r$ from the origin. In other words, it designates a point by means of the angles between a line running from the center of the Earth to the point, the plane of the celestial equator and the plane of a selected initial geodetic meridian. The celestial equator is an imaginary great circle on the sky, lying directly above the Earth's equator. Its plane is perpendicular to the Earth's axis of rotation and contains the Earth's CM. Because the geocentric coordinate system does not agree with the nature, it has not been used in any vehicle location and navigation system. Some literature may use the term geocentric system to describe a reference coordinate system for either a digital map or a GPS receiver, which actually means a geodetic coordinate system with the origin of the reference ellipsoid at the geocenter.

A geodetic coordinate system is a coordinate system consisting of an ellipsoid, the equatorial plane of the ellipsoid, and a meridional plane through the polar axis of the ellipsoid (Figure 2.5). The coordinates of a point in this system are given by the perpendicular distance of the point from the ellipsoid (the ellipsoidal height $h$), by the angle between that perpendicular (the normal) and the equatorial plane (the geodetic latitude $\phi$), and by the dihedral angle between the meridional plane and a plane perpendicular to the equatorial plane and containing the normal (the geodetic longitude $\lambda$). This system designates a point with respect to the reference ellipsoid and with respect to the planes of the geodetic equator and a selected geodetic meridian. The geodetic equator is an ellipse on the reference ellipsoid midway

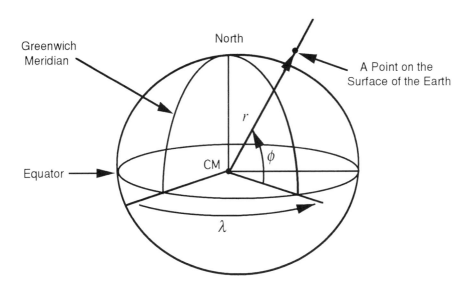

**Figure 2.4** Geocentric coordinate system.

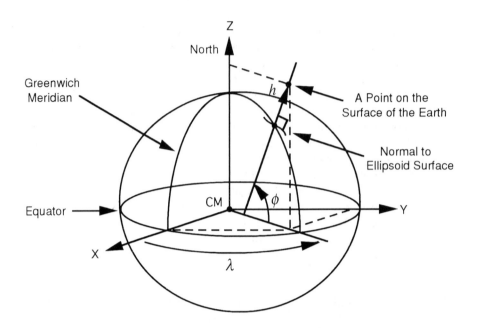

**Figure 2.5** Geodetic and Cartesian coordinate systems.

between its poles of rotation. This equator is the line on which geodetic latitude is 0 degrees, and from which geodetic latitudes are reckoned, north and south, to 90 degrees, at either pole. If the minor axis of the reference ellipsoid is parallel to the rotational axis of the Earth, the geodetic equator will coincide with the Earth's equator.

Three-dimensional (3D) Cartesian coordinates $(X, Y, Z)$ are sometimes used. One such coordinate system is shown in Figure 2.5. The dotted lines indicate the coordinates of a selected point on the Earth's surface. The main advantage of the Cartesian coordinate system is that it is completely defined by the origin, CM in this case, and the three axes. It is much simpler than the other coordinate systems. However, it is not very convenient for location and navigation. For instance, except at the North Pole, moving northeast a certain distance will not result in an identical increase in the $Z$ value. Therefore, this type of coordinate system is rarely used in surface vehicle location and navigation. The geodetic coordinate system referenced to the WGS 84 geocentric ellipsoid (Table 2.1) is commonly known as a conventional terrestrial or Earth-centered-Earth-fixed (ECEF) coordinate system. A Cartesian coordinate system centered at CM can also be considered an ECEF coordinate system. Any point in a geodetic ellipsoidal coordinate system can be translated to a point in Cartesian coordinates provided the origin is the same. Refer to Appendix

A for the exact transformation between 3D Cartesian coordinates and ellipsoidal coordinates.

The following equation describes the relationship between geocentric latitude $\phi_c$ and geodetic latitude $\phi_d$ (Figure 2.6).

$$\tan\phi_c = (1 - e^2)\tan\phi_d$$

From Figures 2.4 and 2.5, we see that the geocentric longitude and geodetic longitude are identical.

Many geodetic coordinate systems are not geocentric, that is, their origins do not coincide with the center of the Earth. This is because offsets may improve the match between the surface of the ellipsoid and the geoid in the relevant region. Latitudes and longitudes referenced to different ellipsoids are different coordinate systems and cannot be mixed. If mixed, this may lead to a displacement of up to 1,500m for a point on the Earth's surface depending on the datum, reference ellipsoid, and the map projection used. This explains why the output of a GPS receiver may not necessarily match a certain map.

Geodetic coordinates are commonly used in modern location and navigation systems. There are three problems with geodetic coordinate systems (as discussed in [1]). First, there are a wide variety of different geodetic systems (Table 2.1). When switching between different systems or integrating portions of a vehicle location and navigation system that use different coordinate system, care must be taken to make ensure that a proper transformation is done. Second, geodetic coordinates are inconvenient for 3D air and space navigation, because the third dimension (height above the reference ellipsoid) provided by this coordinate system is different from the height

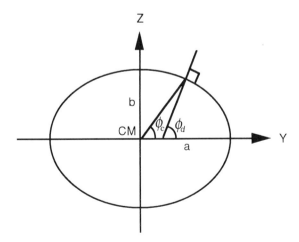

**Figure 2.6** Geocentric and geodetic latitudes from the cross section of the Earth.

above the geoid and other heights relative to different reference ellipsoids. Third, the latitude and longitude cannot be used straightforwardly in a location and navigation system based on geodetic coordinates. Although exact or approximate mathematical formulas can be used to calculate the distance and azimuth, they are not as straightforward as plane trigonometry in 2D Cartesian coordinates. These considerations led to the development of the third method described below, in which the distance and azimuth between two points of known coordinates can be computed easily by plane trigonometry.

To transform geodetic latitudes and longitudes into a planar Cartesian coordinate system, projection equations are used. One valuable projection is the UTM projection. It is often used for map projections involving large countries or continents because a line of constant azimuth is represented by a straight line in this projection.

When integrating a GPS receiver with a dead-reckoned positioning module (Section 3.4) or with a map-matching (Chapter 4) module, one often needs to convert the WGS 84 coordinate system used in the GPS receiver to a UTM coordinate system, so that all relevant observations can be aligned to the surface of the Earth.

$$x = f_1(\phi, \lambda)$$
$$y = f_2(\phi, \lambda)$$

where $x$ and $y$ are the northing and easting in the UTM projection, respectively, and $\phi$ and $\lambda$ are the latitude and longitude in the geodetic coordinate system, respectively.

The UTM projection is coupled with a certain datum that references a geodetic coordinate system. We can transform geodetic coordinates referenced to WGS 84 directly to coordinates in a UTM mapping plane referenced to any ellipsoid. (The 2D transformation equations can be found in Appendix B.) Alternatively, a simple square root equation can be used to calculate the distance between two points in any coordinate system with latitude and longitude as their axes, since the travel area for a particular vehicle location and navigation system is often relatively flat, even though the Earth's surface is curved. The errors of calculation are negligible only if the region is not too big, as is typically the case for most metropolitan areas. However, any such computation must be scaled properly, that is, we must multiply by the appropriate number of meters per degree along each axis. We now discuss the basic principles of the UTM projection.

Representation of the 3D spherical Earth on a 2D mapping plane requires a transformation or projection. Either the angles are preserved (conformal projection, where any angle on the map is exactly the same as on the Earth), or the areas are preserved (equivalent projection, where the ratio between any area on the map and the corresponding area on the Earth is constant). There is no projection that can preserve both.

The Mercator projection is a conformal map projection based on a cylindrical projection. In this projection, the equator is a straight line true to scale, the meridians are straight lines of equal length, and the parallels are straight lines of unequal length. The cylinder is tangent to the Earth along the equator. Turning the cylinder 90 degrees, we have the transverse Mercator projection (Figure 2.7), where the central meridian is a straight line true to scale, and meridians 90 degrees from the central meridian and the equator are also straight lines. Other meridians and parallels are complex curves. Imagine that by wrapping a cylinder around a globe, the cylinder is tangent along a meridian, or secant along lines parallel to this meridian. The meridian and parallel lines are conceptually placed on the cylinder. When the cylinder is cut along a line perpendicular to this meridian, unrolled, and laid flat, the projection is obtained. The ellipsoid (or Earth) is partitioned into zones of equal longitude where the central meridian is in the center of each zone.

The UTM projection is a. specific transverse Mercator projection. First, the ellipsoid is partitioned into 60-degree zones, each of which is 6 degrees, wide in longitude. Second, a scale factor of 0.9996 ($k$ in Appendix B) is applied to the 2D coordinates in order to avoid large distortions in the outer areas of each zone. As noted earlier, UTM coordinates in the north and east directions are referred to as the "northing" and "easting" directions (Figure 2.7). Each zone is divided into smaller squares where the coordinates are given in meters referred to the bottom left boundary of each individual square. In other words, the UTM coordinates are based on a system of zones of longitude, each with its own central meridian. Although the UTM coordinate system makes the life of vehicle location and navigation engineers easier, navigation across UTM zones is difficult. For more information on Earth coordinate systems and map projections, refer to [1–3].

The UTM projection was adopted by the U.S. Army Map Service in 1947 for their use in worldwide mapping. Many of the paper maps used in the United States

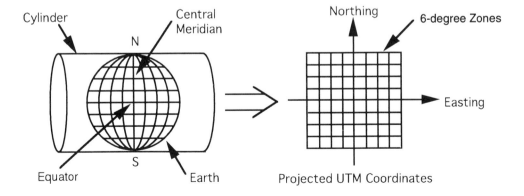

**Figure 2.7** UTM projection.

for local surveying and other mapping operations are also based on transverse Mercator projections. For projection methods used by individual states in the United States, the reader may refer to [4].

There are no widely accepted standards available for digital maps. It is very common for a map database vendor and its customers to agree on a software interface standard. In this way, the interface provides flexibility in the storage and retrieval of data. In the following sections, some of the emerging map database standards are discussed. Two proprietary digital maps are presented as examples of how software interface approaches are used. An introduction to the map compilation process is presented in the last section.

## 2.4 STANDARDS

Recently, the International Organization for Standardization (ISO)/TC 204 has approved a new task on an application program interface (API) for digital map databases, in addition to the existing task to create a map database physical storage format (PSF). Some believe that this will lead to an open global standard in digital map databases. One difficult work lying ahead is to make sure that the industry will actually use these standards once they are developed. The API task group held its first meeting in December 1996. It plans to create a standard method for navigation software to link with different data formats used by different digital map database vendors. The PSF task group aims to standardize the location of data on the storage medium so the system knows where to look for it. In short, the purpose of these activities is to provide a standard mechanism for navigation software to interact with map databases stored on different media as mentioned in Section 2.6.1. These standardization activities will certainly boost the market for vehicle location and navigation systems.

### 2.4.1  Geographic Data Files

The first version of the draft Geographic Data Files (GDF), which is a product of the EUREKA project DEMETER undertaken by Philips and Bosch in Europe, was released in October 1988. This standard was designed to specify the data content, the means by which data representation and structure of data supplied for vehicle navigation systems. The second version of the GDF (GDF 2.0) was developed under work package 3532 of the European Digital Road Map (EDRM) Task Force. The GDF 2.0 Working Group consisted of representatives of EDRM Task Force partners Daimler Benz (prime contractor), Bosch, Philips, Renault, Tele Atlas and Intergraph, and MVA Systematica, a participant in the PANDORA project. The partners in the PANDORA project are Philips, Bosch, the Ordnance Survey, and the Automobile Association. GDF 2.0 differs from GDF 1.0 both in its treatment of the architecture of certain data structures and in its use of terminology. It is also a far more extensive specification. It defines standards for describing, classifying, and encoding features of the road environment, which will support a broad family of application areas.

GDF 2.0 includes the following content: The Feature Catalogue provides a clear definition of "real-world objects" such as roads, buildings, administrative areas, and settlements that are the concern of the broad area of application for the road environment. The Attribute Catalogue defines the characteristics of each feature. Dedicated attributes of particular features are cross referenced. The Relationship Catalogue describes relations between features that can be used to convey information in a realistic manner. The Feature Representation Scheme specifies how real-world objects should be represented in terms of GDF logical structures. The Data Content Specification makes clear which data are required for different applications. The Global Data Catalogue classifies geodetic requirements and other information related to the geometric and geographic characteristics of the data. Finally, the information needed to supply GDF 2.0 data as physical data records is provided in the Media Record Specification.

GDF 2.0 has three major parts: First, the specification of data content (SDC) part lists the information required for navigation systems. Second, the specification of data acquisition (SDA) part describes a set of features, attributes, and relationships for representation of the nonspatial aspects of information. Third, the exchange format (EF) part defines the records and fields.

One important aspect of this standard is the data model in which features, attributes, and relationships are formally defined, along with the concept of a *layer*. Layers are groupings of features that are strongly related such as roads and ferries, administrative areas, settlements, buildings, bridges and tunnels, railroads, and waterways. For example, roads consist of features such as road elements, intersections, and the road itself. In turn, each feature has attributes. For instance, the road element has attributes of traffic direction, name, class, speed limit, impedance value, etc. Relationships represent driving-related information and are a function of several features such as posted sign information and prohibited turns. The geometry and topology of the road network and related geographic features are described in terms of geodetic coordinates, relative location to neighboring features, and shape where appropriate.

After the release of GDF 2.0, GDF 2.1 was released in October 1992, GDF 2.2 was released in November 1994, GDF 3.0 was submitted to Comité Européen de Normalisation (CEN) or the European Committee for Standardization TC 278 for official approval as ENV in October 1995 [5]. The main changes contained in these new versions involve introduction of the concept of a segmented attribute and the concept of time domain, of new information items, and the incorporation of review comments. The revisions and refinements will continue to meet international requirements. There are also several national standards for map transfer, which are moving toward compatibility with the international version of GDF. Some anticipate that the ISO version of GDF will form the foundation for a single international family of transfer standards for ITS map data [6].

GDF is an interchange database standard. Users must develop their own compilers to translate it into *navigable* databases. A navigable database is a database that provides information which can be used to navigate a vehicle through a road network. For instance, in order to provide turn-by-turn route guidance, the database must contain some important attributes such as one-way streets and turn restrictions. Most digital map database suppliers can provide GDF maps for their European customers. Major companies currently involved in the production of map databases are Etak, Inc., Intergraph Corporation, Navigation Technologies Corporation, Tele Atlas International BV, etc.

### 2.4.2 Digital Road Map Association

The Japan Digital Road Map Association (DRM or JDRMA) was established by the Public Works Research Institute of the Japanese Ministry of Construction in 1988 [7,8]. Their product is both a digital map standard and a map database. The association supplies digital road map databases to member companies on magnetic tapes. DRM started digital map database production using 1/50,000 scale topographical maps and is now in the process of refining the database using 1/25,000 scale topographical maps, which offer more detailed and precise features. The topographical maps are issued by the Geological Research Institute and the Ministry of Construction. In addition, they have also used other urban planning maps and information provided by road administrators. They use custom sources, but project all the data onto a reference ellipsoid based on the Tokyo datum. This datum is defined by the location and azimuth on the 1841 Bessel spheroid, with the origin of the datum at the center of the transit circle of the old Tokyo Observatory (Table 2.1).

Three design principles are used by DRM. First, the location and configuration of features on the map are measured on two topographic maps (1:25,000 and 1:50,000) and filed according to a standard regional grid and mesh code system. Second, separate links are established for the road network indicating road configuration, interchanges, and ramps and lanes, with route searching and map matching being integral parts of the system. Third, the digital map consists of two layers of road networks, with the basic road network (main roads) used for the best route computations and the full road network for detailed operations such as map matching and zoom-in graphic displays.

The two layers of the road network are defined as follows. The basic road network consists of the national highways, prefectual highways, and roads 5.5m or wider (two lanes or more). The full road network consists of all roads covered by the basic road network plus all roads 3m or wider (one lane and more). Location is expressed using a mesh code number (with each mesh covering an area of approximately $10 \times 10$ km) and normalized coordinates ($X$: 0-10,000, $Y$: 0-10,000) within each mesh. The road network is defined in terms of a combination of nodes and links (similar to the segments defined earlier). The basic road network uses four-digit node numbers, while the full road network uses five-digit node numbers. Each link number is a combination of the numbers for the nodes at each end. In this way, all roads can be identified by mesh and link numbers. The location and shape of each link is expressed by a series of coordinate values for the interpolating points. The database is organized as a series of files.

DRM has already covered all of Japan and version 3.0 of the map has recently been released. DRM provides the map to their users on magnetic tapes and its data structure is open to public. However, the file format, which is the key to reading the magnetic tapes properly, is only open to DRM users. The DRM does not provide any database interface to users who need to design and implement their own interface to access the map database. Furthermore, DRM does not contain information such as turn restrictions or one-way streets for route planning and guidance. Users need to further expand the database or include smaller scale information (1:10,000) for vehicle navigation.

### 2.4.3 Spatial Data Transfer Standard

The Spatial Data Transfer Standard (SDTS) was begun by the U.S. National Committee for Digital Cartographic Data Standards in 1982. It is designed for transfer of a wide variety of data structures that are used in the spatial sciences. These sciences include cartography, geography, geology, geographic information systems (GIS), and many other related sciences. In 1985, the Standards Working Group of the Federal Interagency Coordinating Committee on Digital Cartography also began work on spatial data exchange standards. During 1987, the results of these parallel efforts were merged by the Digital Cartographic Data Standards Task Force into the proposed Digital Cartographic Data Standards, published as a special issue in 1988 [9].

SDTS is an open and general standard for the exchange of spatial data [10]. It is also a Federal Information Processing Standard (FIPS). Effective February 1994, U.S. federal agencies are required to provide data to other federal agencies and the general public in SDTS format. SDTS consists of three primary parts:

- The first part is the SDTS logical superstructure, which describes the organization and structure of the SDTS transfer mechanism.
- The second part prescribes definitions for spatial features and attributes.
- The third part describes the implementation of the ISO 9211 data transfer standard (i.e., encoding method).

SDTS intends to provide a solution to the problem of transferring spatial (i.e., geographic and cartographic) data from the conceptual level to the details of physical file encoding. Transfer of spatial data involves modeling spatial data concepts, data structures, and the logical and physical structure of the files. To be useful, the data to be transferred must also be meaningful in terms of data content and data quality. SDTS addresses all of these aspects for both vector and raster data structures.

The ITS America (Intelligent Transportation Society of America) Map Database and Information Systems Subcommittee is working on a vector profile for SDTS, which would be a standard for the North American highway database. This project is being worked on in conjunction with the Federal Highway Administration (FHWA).

### 2.4.4 Truth-in-Labeling Standard

A labeling standard, the Truth-in-Labeling Standard, has been completed after several years of work by the Map Database Standards Committee of the Society of Automotive Engineers (SAE) [11]. The purpose of this standard is to define consistent terminology, metrics, and tests for describing and comparing the content and quality of navigable map databases for vehicle location and navigation. The standard will allow users to determine whether a given map database meets the needs of an application. However, it does not specify minimum performance standards or physical database format as the other standards do.

The truth-in-labeling standard contains definitions and detailed descriptions of map database attributes, as well as metrics and tests for each map database object (i.e., physical representation of a concept such as "node" or "segment" in a digital map database) defined. Note that this standard uses "link" in place of "segment." The objects defined include the following:

- Nodes and junctions;
- Location and existence of attached links;
- Road names;
- Address ranges;
- Road classifications;
- Restricted maneuvers.

The present focus of this standard is on in-vehicle navigation applications such as address location, route determination, route guidance, vehicle positioning, and display. Later versions of the standards during the next 3 to 5 years will also be applicable to other ITS map-related applications. In addition, SAE plans to establish a testing institute to test database labels using this standard. Database vendors will specify the content, currency, accuracy, coverage, etc., of their databases in the format specified by the standard. The institute will verify their claims. Based on the verification results, users will be able to quickly and accurately select the database best suited for their applications. As a result, users can determine the usability of a particular database for their applications in advance, rather than finding out later, after the shipment of the database product. In short, the standard allows users with different needs to select the most appropriate database for their applications.

Despite these standards activities, most location and navigation products for automobiles in the United States are based on proprietary digital map databases provided by private companies. The next section covers this topic in detail.

## 2.5 PROPRIETARY DIGITAL MAP DATABASES

### 2.5.1 Etak

The EtakMap databases developed by Etak, Inc., use many sources, and project everything onto a two-dimensional representation of the Earth's reference ellipsoid [12]. The sources used include aerial photography, U.S. Geographical Survey (USGS) maps, local government maps, the U.S. census bureau's TIGER'92 database, cadastral maps, tax assessor information, cross reference directories, ZIP+4 files, field data capture information, customer files, and map reports.

- The USGS produces Quad Sheets, which are accurate 7.5- by 7.5-min contour maps of the United States. Many these maps are 20 to 30 years old and lack highway information.
- TIGER is a standard for vector-encoded data in which features are represented using Cartesian geometry. Features are modeled using combinations of primitives such as points, lines, and areas. TIGER maps were developed by the Census Bureau for taking Census data. They are more complete than USGS maps but they lack accuracy, geometry, and highway data.
- Cadastral maps are maps showing the boundaries of subdivisions of land, usually with bearings and lengths, and the areas of individual tracts, for purposes of describing and recording ownership.

In North America, the reference ellipsoid is based on the North American Datum of 1927 (NAD 27) defined by the location and azimuth on the Clarke spheroid of 1866, with origin at Meades Ranch, Kansas. Clearly, NAD 27 is a geodetic coordinate system. NAD 27 is a horizontal control datum specifying the coordinate system in which coordinates of points on the Earth can be calculated using a set of constants.

As mentioned before, Etak provides interface formats for users to access their map databases. The MapBase format is an ASCII format compatible with the major GIS formats. The MapAccess format is based on Etak 0-track, a proprietary method that relies heavily on the topology of the map, in particular, the carrier topology of the map, rather than geometric coordinates (see Table 2.2). It is said that faster map

**Table 2.2**
Sample Data Structures for Digital Map Databases

| Data Structure | Principle | Usage | Comment |
|---|---|---|---|
| B-trees | Partition data into a hierarchical index tree in which each node occupies a block of memory. | Text indexing | Conventional indexing |
| Carrier-block-trees | Partition data into carrier blocks, each one is the smallest topological closed set. | Spatial indexing | Used by Etak |
| k-d-trees | Partition data recursively into a binary search tree where odd and even levels have different keys to determine branching directions. | Spatial indexing | Popular in GIS but difficult to update compared with quad-trees |
| Oc-trees | Partition data recursively into octants, with possible duplications for features crossing octant boundaries. | Spatial indexing | For 3D data |
| Quad-trees | Partition data recursively into quadrants, with possible duplications for features crossing quad boundaries. | Spatial indexing | Very popular in GIS for 2D data |
| R-trees | Partition data recursively into nested rectangular areas, which may overlap. | Spatial indexing | More flexible than quad-trees or k-d-trees |

data retrieval can be achieved using this method [12]. It is available only by using MapAccess development tools (C language libraries) to retrieve and display maps and geocode addresses. *Geocoding* is the method of applying a latitude and longitude coordinate to an address. Other formats are also available for users with different applications and interface requirements.

EtakMap premium maps are offered in standard and enhanced versions. The enhanced maps will eventually cover 80% of the U.S. population. These maps include address, zip code, road classification, graphic features, political and statistical geography, major landmarks, one-way street encoding for selected urban areas, data overlay attributes, and compliance with National Map Accuracy Standards. In other words, these enhanced maps are navigable. They provide road networks covering more than 43 million metropolitan people. By 1998, Etak plans to have a map database covering major metropolitan areas with a total population of 130 million people. In addition, its standard maps will fill the rural gaps to provide 100% national coverage. These maps include only address and zip code information and are therefore not navigable.

Etak overseas coverage includes Japan, Hong Kong, and Europe, some with complete coverage and some without. Etak has entered into a data-sharing agreement with Tele Atlas BV, a new joint venture formed by Robert Bosch GmbH and Tele Atlas International BV. Etak is focusing its efforts on mapping Great Britain and the United States, and Tele Atlas BV is continuing to map continental Europe, including Switzerland and Austria.

## 2.5.2 Navigation Technologies

The databases developed by Navigational Technologies Corporation (NavTech) also use many sources, and project everything onto a two-dimensional representation of the Earth's reference ellipsoid. The basic database concepts by NavTech contained in NavTech's early interface standards include database content, data organization, data relationships, and data access methodology [13]. Their sources include topographical base maps, aerial photography, and local ground research (field data), as well as information from a variety of additional sources. Each database area is continuously updated and supported by a local field staff and a global customer support organization. Information in the database includes road network geometry, roadway classification, roadway characteristics such as directionality, turn and time-of-day restrictions, cartographic and geopolitical boundaries, and points of interest, landmarks, and services. NavTech databases come with a written warranty of at least 97% accuracy and completeness based on "ground truth." Its guaranteed precisions are within 15m of ground truth in detailed city areas and 100m in areas covered by intertown maps.

In North America, the reference ellipsoid mentioned earlier is based on the North American Datum of 1983 (NAD 83) as defined in the Geodetic Reference System 1980, with its origin at the center of the Earth. From Section 2.3, we know that NAD 83 is a geodetic coordinate system based on a geocentric reference ellipsoid.

The NavTech databases contain information on the road network and environment including roads, intersections, and points of interest [13]. The data are logically organized so that certain information can be accessed and manipulated, and physically organized in such a way as to provide maximum flexibility. The data relationships include a variety of relationships between data entries in the same layer and zone, which means that road names can be superimposed on the correct roadway segment. The data access methodology involves a transparent user interface, which resides between the application and the database itself, thereby allowing users to make high-level logical requests without having to know the structure of the database.

The early NavTech proprietary map-data interchange interface standard, the Standard Interchange File (SIF), consists of a set of record types, in which each record type contains a portion of the database for navigation and digital cartography. Depending on the custom request, either all record types or only a subset of the record types can be created. Record types can be divided into two groups: interchange file (data description directory) records and link records (link information).

Interchange file records contain information about the data format and meaning of each field. Link records contain detailed information about each link. A NavTech link is a specific segment or subsegment in the road network or an area/polygon that is a physical feature or administrative boundary. The interchange file is primarily link oriented and all information about each link is grouped. NavTech requires user applications to connect the endpoints of intersecting links.

In the United States, NavTech has navigable digital databases available for numerous major metropolitan areas. NavTech has completed detailed city coverage for a total area containing more than 100 million people and intertown map databases for less populated areas. These intertown map databases include all roads that are contained on American Automobile Association (AAA) state maps. By the year 2000, NavTech plans to have a map database covering an area with a total population of more than 210 million people.

In Europe, NavTech Europe (formerly European Geographic Technologies (EGT)) has completed a database for more than 80 cities in Germany, the greater Paris area, Lyon, and Marseille in France, Torino and Milano in Italy, and Coventry in the United Kingdom. Additional cities throughout France, Italy, the United Kingdom, and Germany have also been completed. In addition, intertown map databases now include all roads connecting municipalities and settlements larger than $250 \text{ km}^2$.

## 2.6  DIGITAL MAP COMPILATION

As mentioned, a digital map database can be encoded from existing nondigital maps or aerial photographs. The coordinates of nodes are usually encoded using special computers that can record the coordinates of any given point. The conversion or digitizing is done for a point by placing the crosshair of an instrument over the point and pressing a button. In some cases, paper maps are scanned and then converted to vector form by software. Once the basic road geometry has been encoded and the road topology has been checked, the database is further expanded by entering attributes such as road types, road names, speed limits, lengths, turn restrictions, and other details as discussed earlier. Gradually, all these data form an intricate data mesh, a virtual reality of the road network. Although the whole process has been automated to varying degrees, it still requires tedious human involvement and intervention. For instance, one quad sheet for San Antonio, California, took a week to input into a vendor's database. Although a section of the database can be as accurate as the sources, field engineers may still need to actually drive vehicles on the road to verify and update the map for final completion. The final product typically takes the form of raw cartographic data files. The accuracy of these data files is very important to systems replying on map-related functions. Significant errors (greater than 15m in city areas) may lead to poor overall system performance.

Besides providing raw cartographic data files in one of the interchange formats, all the map database vendors also compile their databases in some internal format.

Users can use their map retrieval and drawing libraries to easily access the databases. At this writing, none of the database vendors provides compatible and independent formats for customers. Most location and navigation product companies compile their own navigable databases (files) based on the raw interchange cartographic database files provided by vendors as shown in Figure 2.8. Besides the vendor database files, additional files collected from other sources can also be used as inputs for the map compiler. One example is discussed in Section 6.4.

### 2.6.1 Data Structures

Many different data structures and implementations could be used to store the digital map database [14–16]. Table 2.2 lists several popular data structures, some of which were summarized in [12]. It is not uncommon for a simple map database to use a combination of these and other data structures. The database can be stored either in the main memory of the location and navigation computer or in a secondary storage area such as a hard disk, a CD-ROM, or a PCMCIA memory card (or PC card). Typically, around 10 to 100 MB are required to store an uncompressed metropolitan area map; the size of the map is a function of the data structures selected, the complexity of the attributes, the geographic area covered, the road topology, and other factors. Compression techniques can be used to reduce the size of the map, with the side effect that decompression will be required for applications to access the map database; this often requires additional design considerations and the development of proper mechanisms for dealing with real-time applications. Depending on the system implementation, the user may design a database manager that will provide services to a set of applications requiring access to the database.

The data structures shown in Table 2.2 typically contain, at a minimum, node records and segment records. As an example, the following list contains three possible records and their attributes:

- *Node (point) record:* node ID, latitude, longitude, connectivity, etc.;
- *Segment (line) record:* segment ID, left node, right node, length, speed limit, directionality, address range, road name, city name, state name, zip code, shape points, road type (classification), drivability, etc.;
- *Area (polygon) record:* area ID, area name, city range, city name, state name, zip code, etc.

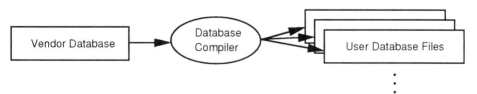

**Figure 2.8** Database compilation.

## 2.6.2  Compiler Structure

The database compiler has a structure identical to that of a conventional language translator. At the top level it consists of two parts: a front end and a back end. The purpose of the front end is to read the vendor data in the interchange format, analyze it, and output the results of this analysis as a set of intermediate files. The analysis process includes determining the connectivity of the road network, finding unique names of objects like streets, cities, countries, points of interest, etc. The purpose of the back end is to take the intermediate files and encode them efficiently as navigable databases (database files) for storage and retrieval by an application or another module. The encoding process includes the creation of B-tree indexes for retrieval of names of streets and cities, the creation of blocks of data containing spatially correlated objects (parcels of land), and the creation of indexes for access to these objects (quad trees, etc.).

If handled properly, the compiler structure and its compilation process should have no effect on map quality. As mentioned earlier, mixed reference coordinate systems in a map-based system can lead to systemic errors as large as 1,500m for every point on the map. On the other hand, errors in the data files themselves will not have such consistency and often occur randomly in several different attributes. For instance, one node on a street might be off 20m of "ground truth," one local street might be missing, a one-way street might be marked as two-way, a particular road might be classified incorrectly, and so on. Such errors frequently occur before compilation of the database and will adversely affect performance of a location and navigation system.

## 2.6.3  Hierarchical Maps

To efficiently solve the complex problem of vehicle location and navigation, the system should at first ignore low-level details and concentrate on the essential features of the problem, filling in the details later. This idea easily leads to generalization, that is, multiple hierarchical layers of abstraction, each focused on a different level of detail. This technique has proven to be very effective in reducing the complexity of large problems (see Section 5.5 for detailed analysis).

As an example, a hierarchical map database and compiler developed by Motorola are briefly discussed. A key feature of this database is that the segments stored in the higher layer are equivalent to a combination of several segments stored in the layer immediately below. Notice that this storage method leads to a considerable reduction in the number of nodes required to represent the exact same road network in a higher layer. As we will learn from the running time analysis of route-planning algorithms in Chapter 5, this reformatting of the road data to create combined road segments for the higher layer results in improvement of the route planning running time because there will be fewer nodes that correspond to segments in the higher layer. The number of nodes to be visited during the planning process is a major factor that affects the running time for the algorithms. An alternative method is to construct a database in which the lower layer segments are reproduced almost exactly in the higher layer. Unlike the previous method, no combination of the low-layer segments is required. Because this method simply removes certain local roads when moving up to the higher layer, the remaining road segment data will be reproduced exactly from the lower layer to the higher layer. This will require more memory, but reduces the compilation time and makes the database design less complicated. However, route planning is relatively time consuming using this type of database, because there is no substantial reduction in the number of nodes visited during the decision-making process.

The Motorola hierarchical map database is a collection of layers. The full set consists of all four layers, layer 0 (the bottom layer), layer 1, layer 2, and layer 3.

- Layer 0 includes all the roads in the network and related navigation information.
- Layer 1 includes collector roads, arterials, and highways.
- Layer 2 includes arterials and highways.
- Layer 3 includes highways only.

Collector roads are semi-major roads, which may lead to arterials. Although a four-layer map is used, the database compiler can support up to eight layers. A graphical representation of a four-layer sample map is illustrated in Figure 2.9. It is possible to consider maps that are composed of a single layer, such as layer 2 only, or multiple layers such as layers 1 and 2. Such maps would cover large areas and contain fewer segments, thus requiring less storage space than fully populated maps. These maps are direct analogs of the country and state maps that we use when traveling from city to city.

As discussed before, the database compiler, with its front-end and back-end structure, compiles the raw cartographic data into navigable hierarchical data files that include map parcel files and index files. These files can then be stored on CD-ROM as shown in Figure 2.10, for access by other modules via the application interface (API). A parcel is analogous to a portion of a map area, which is a collection of geographical objects (such as segments) falling within a rectilinear region defined by minimum and maximum latitudes and longitudes. The map index data help to locate multiple maps stored on the CD-ROM (such as maps for different states). The map parcel data contain detailed information on the map for each area. Index data are used to locate the parcel which stores vector-encoded map representations. A similar CD-ROM storage method has been used by Philips [17] and others [18,19].

Another feature of the Motorola hierarchical database is the addition of a dynamic database (see Figure 2.10). The dynamic data are used to store dynamically updated information for each *link* received from a traffic information center. In this

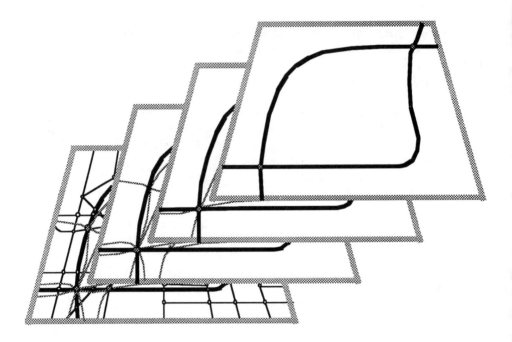

**Figure 2.9** Hierarchical map database.

terminology, link refers to a segment with turning delay information. A turning delay is the normal time required for a vehicle to finish a turn associated with a particular segment. Note that this definition of a link is different from the one used in our discussion of the DRM and NavTech databases. Depending on the number of turns ahead, a specific segment may need more than one link to distinguish the delays for different turns. The database module and its application modules can incorporate these real-time updates for each segment and use the turn delays for specific applications. Applications for the dynamic database are discussed in further detail in Section 6.4.

One of the key benefits for developing a hierarchical map database is to support hierarchical route-planning algorithms in an environment where a slow secondary database storage device such as a CD-ROM is being used. All modules work at layer 0, except for the route-planning module. The database has detailed data at layer 0 and minimal data at other layers. In this way, for route planning, the system can fetch data that cover a larger area during a single secondary-memory read for the higher layer than for the bottom layer. Section 5.5 will contain an additional discussion of the benefits of hierarchical map databases for route planning. Another application of hierarchical storage is in map display. Depending on the scale of the map requested by the user, the map display application will select the appropriate

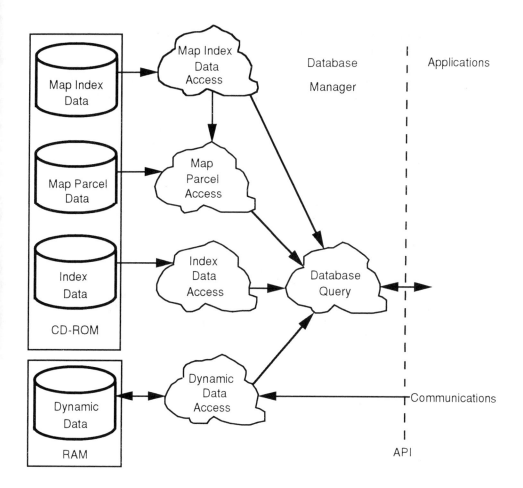

**Figure 2.10** Database access.

map layer needed to draw the map. This will reduce the amount of data read from the CD-ROM and reduce the processing time required to filter unwanted information.

In summary, digital map database design should take into account the following four considerations: content, organization, relationships, and access methodology. In other words, a vehicle location and navigation system needs a map database that can support map matching, route planning, route guidance, map display, and the display of associated points of interest. To compile a vendor database into a customized database (data files), database compiler designers must consider the specific application domains involved. For vehicle navigation systems, a navigable database should at a minimum provide accurate map-related location information, the attributes required to calculate a best route and to guide the vehicle, as well as road-related facilities and information. The database must be easy to access. For an on-board navigable database, the run-time access speed must be carefully addressed.

Before turning to the next chapter, let us consider how street addresses can be correlated with digital map coordinates (geocoding) assuming that the map does not contain the coordinates of every possible address. Given the map database concepts we learned in this chapter, how can we derive the coordinates for a known address?

## References

[1] V. Ashkenazi, "Coordinate Systems: How to Get Your Position Very Precise and Completely Wrong," *The Journal of Navigation*, Royal Institute of Navigation, Vol. 39, No. 2, May 1986, pp. 269–278.

[2] L. M. Bugayevskiy and J. P. Snyder, *Map Projections: A Reference Manual*, Bristol, PA: Taylor & Francis, 1995.

[3] J. P. Snyder, *Map Projections—A Working Manual*, U.S. Geological Survey, 1987.

[4] A. Leick, *GPS Satellite Surveying*, 2nd ed., New York: John Wiley & Sons, 1995.

[5] European Committee for Standardization (CEN), *Geographic Data Files*, Brussels: CEN, Oct. 1995.

[6] R. Pearlman and S. Scott, "IVHS Map Database Transfer Standards: Current Status," *Proc. IEEE-IEE Vehicle Navigation and Information Systems Conference (VNIS '95)*, July 1995, pp. 368–373.

[7] Japan Digital Road Map Association, *Digital Road Map Database Standards*, Version 2.0, Jan. 1991.

[8] M. Shibata and Y. Fujita, "Current Status and Future Plans for Digital Map Database in Japan," *Proc. IEEE-IEE Vehicle Navigation and Information Systems Conference (VNIS '93)*, Oct. 1993, pp. 29–33.

[9] J. Morrison, "Proposed Standard for Digital Cartographic Data," *The American Cartographer*, Vol. 15, No. 1, Jan. 1988, pp. 9–140.

[10] U.S. Geological Survey, *Spatial Data Transfer Standard*, 1992.

[11] Society of Automotive Engineers, *Truth-in-Labeling Standard for Navigation Map Databases*, SAE J1663, Aug. 1995.

[12] M. S. White, "Digital Maps—A Fundamental Element of IVHS," *IVHS Journal*, Vol. 1, No. 2, 1993, pp. 135–150.

[13] D. C. Marsh, "Database Design, Development, and Access Consideration for Automotive Navigation," *Proc. IEEE-IEE Vehicle Navigation and Information Systems Conference (VNIS '89)*, Oct. 1989, pp. 337–340.

[14] O. Gunther, *Efficient Structures for Geometric Data Management*, Berlin: Springer-Verlag, 1988.

[15] H. Samet, *The Design and Analysis of Spatial Data Structures*, Reading, MA: Addison-Wesley, 1990.

[16] M. S. White and G. E. Loughmiller, "Apparatus Storing A Representation of Topological Structures and Methods of Building and Searching the Representation," *Canada Patent No. 1277043*, Nov. 1990.

[17] M. L. G. Thoone, "CARIN, A Car Information and Navigation System," *Philips Technical Review*, Vol. 43, No. 11/12, Dec. 1987, pp. 317–329.

[18] H. J. G. M. Benning, "Digital Maps on Compact Disc," *SAE Paper No. 860125*, Society of Automotive Engineers, 1986.

[19] K. Ishikawa, M. Ogawa, T. Tsujibayashi, and Y. Shoji, "Digital Map on CD (Called CD Information)," *SAE Paper No. 880221*, Society of Automotive Engineers, 1988.

# CHAPTER 3

▼▼▼

# POSITIONING MODULE

## 3.1 INTRODUCTION

The positioning module is a vital component of any vehicle location and navigation system. To either help users obtain vehicle location or provide users with proper maneuver instructions, the vehicle location must be determined precisely. Therefore, accurate and reliable vehicle positioning is an essential prerequisite for any good vehicle location and navigation system.

Positioning involves determination of the coordinates of a vehicle on the surface of the Earth. Location involves the placement of the vehicle relative to landmarks or other terrain features such as roads. These tasks are accomplished by the positioning module in conjunction with other modules. In this chapter, we introduce the basic positioning technologies and sensors and then study sensor fusion technologies for integrating the sensor inputs used to estimate position and location.

Three positioning technologies are most commonly used: stand-alone, satellite based, and terrestrial radio based. Dead reckoning is a typical stand-alone technology. A common satellite-based technology involves equipping a vehicle with a global positioning system (GPS) receiver. Dead reckoning and GPS technologies have been used widely in vehicles. Therefore, we will concentrate on these two commonly used and easily accessed technologies in this chapter and leave the terrestrial radio technology to a later section (Section 9.2) following our discussion of radio communications in Chapter 8. Although some of these radio technologies, like GPS, do not require users to construct central host facilities or complicated communications systems and infrastructures, this arrangement will quickly direct readers' attentions to the more popular technologies first. Readers may refer to Table 10.1 for a performance comparison of the main positioning technologies discussed in the book.

As we will learn shortly, no single sensor is adequate to provide position and location information to the accuracy often required by a location and navigation system. The common solution (frequently the only way of obtaining the required levels of reliability and accuracy) is to fuse information from a number of different sensors, each with different capabilities and independent failure modes. Therefore, a positioning module typically integrates multiple sensors, which compensate for one another to meet overall system requirements. This in turn requires that we study a variety of sensors (Figure 3.1), fusion methods, and algorithms. Because a GPS receiver functions like an absolute sensor, for convenience of study we discuss it together with the compass. Bear in mind that this is a satellite-based technology and is based on an operating principle quite different from that of other sensors (as we will learn shortly).

The positioning module is the most fundamental of the various modules needed for vehicle location and navigation. As seen from Figure 3.1, it is based on a variety of different positioning sensors. Because of the key roles played by these sensors, we will discuss commonly used low-cost sensors in depth. Readers who are not interested in the details can skip the in-depth discussions but still should be able to grasp the basic concepts in this chapter. We do not pretend to cover all possible sensors in this chapter. With the rapid advances of technology, it is difficult to cover even a particular type of sensor completely in a limited space since a variety of implementation methods can be used to manufacture a sensor based on the same theoretical principle. A similar difficulty also occurs with sensor integration and fusion technologies. However, the emphasis placed on sensors in current use and

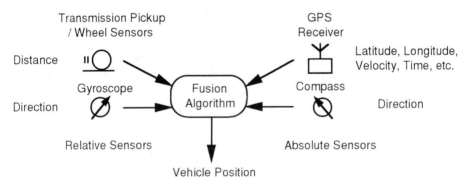

**Figure 3.1** A generic positioning module.

the fundamental fusion technologies available should be sufficient for anyone interested in vehicle location and navigation to have a basic understanding of this particular area. For available linear and angular automotive sensor technologies, see the two tables summarized in Appendix C. Reviews of the latest automotive sensor development and automotive sensor technologies can be found in [1–4]. A detailed discussion of sensor technologies, sensor principles, and sensor interface circuits can be found in [5,6]. In-depth coverage of sensor integration and fusion can be found in [7].

## 3.2 DEAD RECKONING

A very primitive positioning technique is dead reckoning. For vehicles traveling in a two-dimensional planar space, it is possible to calculate the vehicle position at any instance provided that starting location and all previous displacements are known. Dead reckoning incrementally integrates the distance traveled and direction of travel relative to a known location. In short, it is a technique that determines the vehicle location (or coordinates) relative to a reference point.

The sensors discussed in the following sections can be used to measure the direction of the vehicle $\theta$ and the distance traveled $d$. After the sensor data have been sampled and integrated and sensor fusion has been performed, the vehicle position $(x_n, y_n)$ and orientation $(\theta_n)$ at time $t_n$ can be calculated from the equation

$$x_n = x_0 + \sum_{i=0}^{n-1} d_i \cos \theta_i$$

$$y_n = y_0 + \sum_{i=0}^{n-1} d_i \sin \theta_i$$

$$\theta_n = \sum_{i=0}^{n-1} \omega_i$$

where $(x_0, y_0)$ is the initial vehicle position at time $t_0$, $d_i$ is the distance traveled or the magnitude of the displacement between time $t_{n-1}$ and time $t_n$, $\theta_i$ is the direction (heading) of the displacement vector, and $\omega_i$ is the angular velocity for the same time period. A drawing outlining this method is shown in Figure 3.2.

When the sampling period is constant (and short relative to the time scale for changes in vehicle velocity), the above equation can be written as

$$x_n = x_0 + \sum_{i=0}^{n-1} v_i T \cos(\theta_i + \omega_i T)$$

$$y_n = y_0 + \sum_{i=0}^{n-1} v_i T \sin(\theta_i + \omega_i T)$$

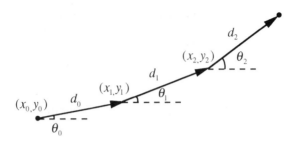

**Figure 3.2** Dead-reckoning method.

where $v_i$ is the velocity measured over the sampling period $T$. During each sampling period, $v_i$ and $\omega_i$ are essentially constant. Mathematically speaking, this method basically amounts to a continuous integration of successive displacement vectors.

As seen from the preceding equations, the current calculated vehicle position during each sampling period (cycle) depends on the previous calculation cycle. It is difficult to eliminate errors associated with the previous cycle or the current measurement (due to sensor inaccuracy and the assumption that the heading remains constant over the sampling period). If not compensated or not properly compensated, these errors will generally accumulate as the vehicle continues to travel, and the calculated position of the vehicle will become less and less accurate. We will discuss algorithms (Chapter 4), short-range beacon networks (Section 8.3.7), and other technologies (Section 9.2) for eliminating these cumulative errors.

Position sensors can be designed to detect various components (distance, direction, or angular velocity) of the position of vehicular mechanical systems. They are either directly coupled to a shaft or linkage, or indirectly coupled to these or other vehicle parts in the case of a noncontact or proximity sensor. Many factors affect the choice of a sensor for a particular situation in the vehicle design, such as frequency of vibrations, temperature range, or the presence of dirt and grease. For a designer of a location and navigation system, position sensors and their output signals are often already given. The task is to make them function as an integral part of the positioning module. Therefore, a basic understanding of common sensor technologies becomes necessary.

## 3.3 RELATIVE SENSORS

A relative sensor is a device that can measure the change in distance, position, or heading based on a predetermined or previous measurement. Without knowing an initial position (or previous reference) or heading, this sensor cannot be used to determine absolute position or heading with respect to the Earth.

### 3.3.1  Transmission Pickups

Transmission pickup sensors are used to measure the angular position of the transmission shaft. Different technologies, such as variable reluctance, the Hall effect, magnetoresistance, and optically based technologies, are used to convert the mechanical motion into electronic signals. Although some transmission pickup sensors may have analog output, it is not difficult to design an interface to convert the signal to pulses indicating fractional revolution. Knowing the number of pulse counts per revolution and a proper conversion scale factor, one can convert the output of the sensor into distance traveled. To avoid mechanical wear and corresponding changes in accuracy of the measurement, it is desirable to measure the angular position of the shaft with a noncontact sensor. Because there is always an air gap between the sensor and the component being monitored, a noncontact sensor is not subject to friction wear. Magnetic fields and optical methods are the two most common methods used for noncontact coupling to a rotating shaft. In this section, we discuss only the magnetic field method.

There are two common places in the vehicle where the signal from the transmission pickup sensor can be obtained. One is from a connector behind the speedometer. The other is from a connector on the transmission housing. In general, a two-wire connector indicates a variable-reluctance position sensor and a three-wire connection indicates a Hall-effect position sensor.

These variable-reluctance and Hall-effect sensors are not limited to transmission pickups. They can be used to detect the position and speed of rotating toothed or notched wheels in crankshaft-, camshaft-, and wheel-monitoring applications. Some of these applications are discussed in the next section.

### *Variable-Reluctance Position Sensor*

The principle of the variable-reluctance sensor is that the total electromagnetic force (emf) induced in a closed circuit is equal to the time rate of decrease in the total *magnetic flux* linking the circuit. The total magnetic flux (number of lines of magnetic force) is inversely proportional to total reluctance. Any change in the reluctance of the magnetic circuit will cause a change in the magnetic flux. If the changing flux lines pass through a coil, the change in flux will induce an emf on the coil. The output emf is measured as a voltage proportional to the rate of change of the reluctance for the path as shown in the following equation:

$$e = -\frac{d\Phi}{dt}$$

where $d\Phi/dt$ is the rate of change in the magnetic flux in webers per second, and $e$ is the emf in volts. As a result, the output voltage goes to zero as the rate of change approaches zero. Therefore, variable-reluctance sensors cannot be used at speeds near zero. Additional detail on magnetic field measurements and magnetic flux is provided in Section 3.4.1.

The variable-reluctance position sensor consists of a permanent magnet with a coil of wire wound around it (Figure 3.3). A ferrous wheel or iron/steel disk mounted on the rotating shaft has teeth or tabs that pass between the pole pieces of this magnet. The voltage output from the sensor is typically a sine wave. When a tooth on the ferrous wheel begins to pass between the magnet pole pieces, the coil voltage starts increasing from zero, reaches a maximum, and falls to zero again when the tooth is exactly between the pole pieces. At this point the rate of change in the magnetic flux is zero (even though the magnetic flux is at its maximum value). Therefore, the induced voltage across the sensing coil goes to zero. As the tooth continues its movement, the voltage output from the coil once again increases from zero, reaches a maximum, and falls to zero again as the tooth passes out of the gap between the pole pieces, and thus finishes a complete sine cycle.

Despite their simple, rugged construction and low cost, variable-reluctance sensors do have drawbacks [2,8]. First, it is difficult to output a constant voltage envelope. As the speed of the ferrous wheel changes, the voltage level and frequency change as well. Second, the signal-to-noise ratio can be seriously degraded by vibrations or resonance. Third, the output of the sensor is sensitive to the gap between the sensor and the disk. In other words, the output is inversely proportional to the gap size. Proper alignment and a rigid sensor mounting are required. Fourth, there is no sensor output when the disk rotation speed is below a certain threshold. This means that no output is available at very low speeds. Fifth, the range of this sensor tends to flatten at high speeds. The extent of this effect varies based on the sensor design. Finally, electromagnetic interference or radio-frequency interference can

Side
View

Front View of
Ferrous Wheel

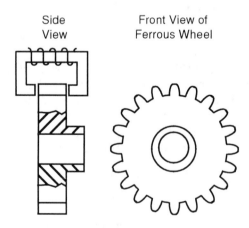

Figure 3.3 Variable-reluctance position sensor.

introduce false signals if the sensor is not properly shielded or packaged, since the coil in the reluctance sensor has a rather high impedance. Signal processing circuitry can be added to the sensor package to obtain a high signal level and to minimize the interference.

*Hall-Effect Position Sensor*

The Hall effect is the effect by which a voltage is generated in a conductor moving in a magnetic field. It is based on the interaction between moving electric carriers in a conductor and an external magnetic field. As a conductor is moved through a magnetic field, the current (which consists of moving electric charges) generated in the conductor will experience a force in a direction perpendicular to both the direction of motion and the magnetic field. A voltage due to the movement of these charges can be detected. Hall-effect devices can be manufactured in silicon or semiconductor materials.

A Hall-effect position sensor with a ferrous wheel or iron/steel disks is shown in Figure 3.4. The Hall-effect device is assembled into a probe with a biasing magnet and a circuit board. Figures 3.4 does not show the configuration inside the probe. The Hall-effect device actually lies at the end of the probe, close to the wheel, so that it always lies between the magnet and the wheel. When an external current is passed through this device, a voltage develops across the device perpendicular to the direction of current flow and to the direction of magnetic flux. As the ferrous wheel rotates, the reluctance of the magnetic field changes as the teeth or tabs pass the probe.

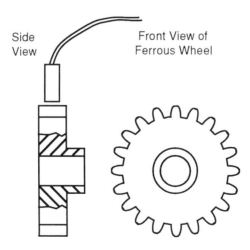

Figure 3.4  Hall-effect position sensor.

Unlike the variable-reluctance sensor, the Hall-effect sensor is a zero-speed sensor. No matter what the velocity, once the Hall-effect device lines up with the strongest magnetic field, there will be a maximum waveform output. Therefore, the voltage output (square wave) from this sensor is independent of the speed of the rotating shaft.

Sensors that utilize Hall-effect technology also have some limitations. The response degrades as the upper and lower limits of operating temperature are approached. They cannot be used in severe temperatures, lower than −40°C (−40°F) or higher than 65.6°C (150°F). The presence of other close, strong magnetic fields can cause errors. As in the case of the reluctance sensor, Hall-effect sensors require careful alignment and mounting because of sensitivity to the gap size between sensor and disk. Unlike the passive reluctance sensor, the Hall-effect sensor is an active sensor which requires an external electrical input. In addition, unlike the variable reluctance sensor, the output voltage of the Hall-effect sensor remains constant as the disk rotation frequency is varied. This sensor can easily be combined with a signal processing circuit to output a high-level voltage. In general, it is more desirable to use a Hall-effect sensor than a variable-reluctance sensor for transmission pickup sensors or other position sensors used in vehicle location and navigation systems. Further discussion of these and other automotive sensors can be found in [2,4].

### 3.3.2  Wheel Sensors

*Wheel Sensors for Anti-Lock Braking Systems*

A typical anti-lock braking system (ABS) consists of wheel speed sensors, an electronic control unit, a brake pressure modulator, wiring, relays, hydraulic tubing, and connectors to link the whole system together. Most ABS systems currently in use offer four-wheel control. The basic components used for the wheel speed measuring systems include a sensing element, a ferrous wheel (exciter wheel, target wheel, tone wheel, or iron/steel disk), and the mechanical hardware required to mount the above elements with the appropriate spacings to ensure correct operation.

The variable-reluctance sensor is still the most popular for ABSs because of its low cost and relative reliability. As discussed earlier, this sensor is not reliable at low speeds, especially below 0.45 to 1.34 m/s (1 to 3 miles/hr); this makes accurate vehicle positioning very difficult when using these sensors for input to a positioning module. Hall-effect position sensors can also be used as ABS speed sensors. We have learned that both sensors are electromagnetic pulse pickups using toothed wheels (exciter rings) mounted directly on the rotating components of the drive train, shaft, or wheel hubs. They provide a digital signal whose frequency is proportional to the rotational velocity of the wheel. Furthermore, any technology with a sensing element that converts mechanical motion into an electrical signal can also be used as a wheel sensor. Table 3.1 compares several different wheel sensor technologies (most of the items in this table are from [9]). From Table 3.3 we notice that only the variable-reluctance sensor is fundamentally incapable of measuring at zero speed (another exception is that a Hall-effect sensor cannot function as a zero-speed sensor if its feedback circuit has a capacitor).

## Differential Odometer

Differential odometry is a technique to provide both distance traveled and heading change information by integrating the outputs from two odometers, one each for a pair of front or rear wheels. An odometer is a relative sensor that measures distance traveled with respect to an initial position. A wheel odometer typically measures the number of rotation counts (pulses) generated by a rotating wheel. Averaging the left and right wheel rotation counts and multiplying by a proper scale factor enables the distance traveled by the vehicle to be determined. The difference between the counts for the left and right wheels multiplied by the same scale factor and divided by the axle length can be used to obtain the change in the heading of the vehicle. Both the distance traveled and the heading change can be used in calculating the vehicle position.

One example of a differential odometer is shown in Figure 3.5. A ferrous wheel or iron/steel disk (toothed gear) is attached to each of the nondriven wheels of the vehicle (because acceleration and deceleration have less effect on the output of sensors on nondriven wheels). A Hall-effect sensor is located close to the toothed gear. This sensor outputs a pulse for each tooth that passes across its sensing tip. Two counters, one for the left wheel and one for the right wheel, count these pulses. This left or right tooth count can be translated to the distance each wheel has traveled by multiplying the count by the scale factor (which is a predetermined value for the distance per count).

**Table 3.1**
Wheel Speed Sensor Comparison

| Sensor | Frequency Range | Magnet | Ferrous Wheel | Signal/Noise |
|---|---|---|---|---|
| Eddy current | 0–500 kHz | No | No | High |
| Hall effect | 0–1 MHz | Yes | Yes | Moderate |
| Magnetic transistor | 0–500 kHz | Yes | Yes | Moderate/high |
| Magnetoresistive | 0–5 MHz | Yes | Yes | Moderate |
| Optical | 0–10 MHz | No | No | Very high |
| Reed switch | 0–600 Hz | Yes | Yes | High |
| Variable reluctance | 1–100 kHz | Yes | Yes | Low |
| Wiegand effect | 0–20 kHz | Yes | Yes | High |

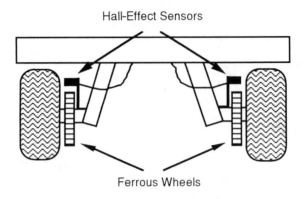

**Figure 3.5** Differential odometer.

After resetting the counters and driving the vehicle along a known straight distance $D$ while recording the counter numbers, the left and right scale factors can be calculated from

$$K_L = \frac{D}{C_L}$$

$$K_R = \frac{D}{C_R}$$

where $C_L$ and $C_R$ are the number of counts for the left wheel and right wheel, respectively. Knowing the scale factor for each wheel and the previous measurement at $t - 1$, we can easily derive the distance traveled up to time $t$ as follows:

$$d(t) = d(t - 1) + \frac{K_L C_L + K_R C_R}{2}$$

When the vehicle heading is changed, the outer wheel travels farther than the inner wheel by $|K_L C_L - K_R C_R|$. Knowing the distance $L$ between the left and right wheels (axle length), the vehicle heading can be obtained from this equation:

$$\theta(t) = \theta(t - 1) + \frac{|K_L C_L - K_R C_R|}{L}$$

Note that as long as the ratio $K_L/K_R \approx 1$, we know that the vehicle is traveling on a straight-line segment road (provided $K_L$ and $K_R$ are accurate). On the other hand, if we know that the vehicle is traveling on a straight-line segment based on a given digital map, we can use this information to calibrate these two scale factors dynamically. Moreover, the outputs from other sensors (such as GPS, see Section 3.4.2) can be used to calibrate the scale factor. For instance, when the vehicle travels above a certain speed and the tracked satellite geometry used for the position fix is good, the velocity and heading information received from the GPS sensor is very good. The travel distance extracted from the velocity can then be used to calibrate the scale factor. Now we can see the benefit of using multiple positioning sensors in a location and navigation system.

Depending on the technology used, an odometer or differential odometer often has an operational error similar to that experienced by variable-reluctance and Hall-effect sensors in general. In addition, the differential odometer is subject to systematic and nonsystematic errors such as unequal wheel diameter, misalignment of wheels, the finite sampling rate and count resolution, running over objects on road, slips or skids involving one or more wheels, etc. A slip or skid occurs if the vehicle accelerates or decelerates too rapidly or travels on a snowy/icy/wet road. These errors may lead to incorrect calculated distance and heading data. Furthermore, in sharp turns, the contact point between each wheel and the road can change, so that the actual distance $L$ between the left and right wheels will be different from the one used in the preceding equation to derive the heading. The tire pressure in each wheel may change over time, as may the wheel diameter due to the tire conditions or change in temperature. Therefore, the scale factors used to calculate position may change after initial calibration. These factors contribute to the error sources of the differential odometer, which lead to accumulative errors discussed in Section 3.2.

### 3.3.3 Gyroscopes

Rate-sensing gyroscopes measure angular rate, and rate-integrating gyroscopes measure attitude. At the present time, most location and navigation systems use gyroscopes to measure the angular rate.

Gyroscopes include mechanical, optical, pneumatic, and vibration devices. A comparison of these gyroscope types is presented in Table 3.2. (The gas-rate gyroscope is classified as a pneumatic gyroscope for lack of a better word.) As the explosive development of design and manufacturing technologies continues, the

**Table 3.2**
Gyroscope Comparison

| Type | Performance | Cost | Size |
|------|-------------|------|------|
| Mechanical | Very good | Expensive | Large |
| Optical | Very good | Moderate | Moderate |
| Pneumatic | Good | Moderate | Moderate |
| Vibration | Good | Cheap | Small |

characteristics of each type of gyroscope in Table 3.2 will improve over the years, and other new technologies may mature to the point where they can be used to further expand this table.

The vibration gyroscope is currently the device of choice for many vehicle location and navigation systems. Discussions of other techniques can be found in [5,10,11]. There are four different types of vibration gyroscope based on the bar (or beam) shape: tuning fork, rectangle, triangle, and cylinder as shown in Figure 3.6. A small electronic circuit is commonly used to control the vibration, convert measurements to an angular velocity output, and provide an interface to the user.

A vibration gyroscope measures the Coriolis acceleration—a phenomenon discovered by G. G. de Coriolis. An object moving perpendicular to a rotating frame will cause an observer on the rotating frame to see an apparent acceleration of the object. The Coriolis acceleration is generated by the angular rotation of a vibrating bar or fork. A change in the direction around one axis of a driving element (transducer) induces a vibration in a detection element (transducer) on another axis. Vibrating piezoelectric ceramics are often used as transducers in a vibration gyroscope and are attached to the bars shown in Figure 3.6.

In general, the apparent Coriolis acceleration $a$ is given by

$$a = 2\omega v \sin\phi$$

where $\omega$ is the angular velocity of the rotating axes relative to a fixed set of axes, $v$ is the velocity of the bar, and $\phi$ is the angle between the vectors $\omega$ and $v$. The

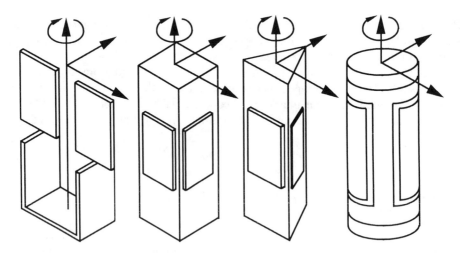

**Figure 3.6** Vibration gyroscopes.

acceleration is at right angles to both $\omega$ and $v$. Assuming the bar has mass $m$ and $\sin \phi = 1$, we derive the Coriolis force as

$$F = 2m\omega v$$

Notice that the Coriolis force is proportional to $\omega$ in the preceding equation. This Coriolis force is detected and converted to a voltage by the detector element and a simple circuit. Since the mass of the bar and the vibrational velocity $v$ of the bar excited by the driving element (and the associated oscillator circuit) are both known quantities, the angular velocity $\omega$ may easily be derived and provided to the user as the output of the gyroscope. This angular velocity must be sampled and integrated to produce the change in the vehicle direction $\theta$:

$$\theta = \int \omega dt$$

Gyroscope performance is measured by a variety of factors. Scale factor and drift (bias and offset) are major parameters used to characterize gyroscope accuracy. The scale factor describes the capability of the gyroscope to accurately sense angular velocity at different angular rates. This value is used to convert the voltage output from the gyroscope into an angular rate or direction. System imperfections can cause small variations in the scale factor. These variations appear as a factor at the output of the gyroscope. Gyroscope drift characterizes the ability of the gyroscope to reference all rate measurements to the nominal zero point. It appears as an additive term on the gyroscope output, so any actual drift away from a given or predetermined value will cause the angular errors to accumulate. The gyroscope drift model varies considerably between different sensor designs. For instance, the following error model, with adjustable parameters $C_1$, $C_2$, and $T$, has been suggested for a triangle-bar vibration gyroscope [12] ($t$ is the time since initialization):

$$\epsilon(t) = C_1(1 - e^{-t/T}) + C_2$$

Recommended methods for reducing the gyroscope errors caused by temperature or other factors include using an analog-to-digital (A/D) converter and filters and measuring the offset voltage at zero angular velocity, just before the output of the sample-and-hold circuit [13,14].

In addition to the methods mentioned, users can also build compensatory models to dynamically calibrate a gyroscope. For instance, one method of calibrating the scale factor involves using other sensors to sense the angular rate and modify the scale factor. Another approach involves recalibrating the scale factor whenever a known turn maneuver occurs based on the digital map or other sensors. One method of calibrating the drift involves calibrating the predetermined or manufacturer-given value whenever the vehicle is stationary. Another method is to correct the value whenever the vehicle is not turning, as determined by other sensors or the map-matching algorithm (Chapter 4).

## 3.4  ABSOLUTE SENSORS

Absolute heading and position sensors are very important in solving location and navigation problems. A relative sensor alone cannot provide an absolute direction or position with respect to a reference coordinate system. In contrast, an absolute sensor can provide information on the position of the vehicle with respect to the Earth. The most commonly used technologies for providing absolute position information are the magnetic compass and GPS.

### 3.4.1  Magnetic Compasses

A magnetic compass measures the Earth's magnetic field. When used in a positioning system, a compass measures the orientation of an object (such as a vehicle) to which the compass is attached. The orientation is measured with respect to magnetic north. Magnetic field sensors are also known as magnetometers. The intensity of a magnetic field can be measured by the magnetic flux density $B$. The units of measurement are the tesla (T), the gauss (G), and the gamma ($\gamma$), where 1 tesla = $10^4$ gauss = $10^9$ gamma.

The Earth's magnetic field has an average strength of 0.5 G. It can be represented by a dipole that fluctuates in both time and space. The best-fit dipole location is approximately 440 km off center and is inclined approximately 11 degrees to the Earth's rotational axis [5]. The difference between true north and magnetic north is known as *declination*. This difference varies with both time and geographical location. To correct the difference, declination tables are provided on maps or charts for any given locale. For a vehicle location and navigation system, this does not cause a problem as long as all directions used in the system are referred to the same (magnetic or geographic) north.

Because of portability, high vibration durability, and quick response, the electronic compasses have many advantages over conventional compasses. Among the electronic compasses, the fluxgate compass is the most popular one. The sensitivity range of this sensor is from $10^{-6}$ G to 100 G and the maximum frequency is about 10 kHz [15]. The term *fluxgate* is obtained from the gating action imposed by an ac-driven excitation coil that induces a time-varying *permeability* in the sensor core. Permeability is the property of a magnetizable substance that determines the degree to which it modifies the magnetic flux of the magnetic field in the region occupied by the magnetizable substance. This property in the magnetic field is analogous to conductivity in an electrical circuit. The absolute permeability $\mu$ of a given material is a measure of how well it serves as a path for magnetic flux. A magnetic field force *H* applied to a high-permeability material induces a magnetic flux density

$$B = \mu H$$

where the magnetic flux density *B* is analogous to the current density in an electrical circuit. A magnetization curve can be plotted from the preceding equation, where the permeability $\mu$ is the slope (Figure 3.7).

High-permeability materials have two interesting properties: hysteresis and saturation. Hysteresis refers to the fact that the magnetic flux through the material lags behind the change in the magnetic field. Due to hysteresis, $\mu$ depends both on the current value of *H*, and the history of previous values for *H*, as well as the sign of *dH/dt*. This causes the curve to follow a different path as *B* increases and decreases to form a hysteresis loop (Figure 3.7). Materials saturate (in other words, further increases in magnetic flux become impossible) at high magnetic flux densities. If a material such as a core is forced into saturation by some additional magnetizing force, the flux lines of the external field will be relatively unaffected by the presence of this saturated core. The fluxgate compass uses the phenomenon of core saturation to directly measure the strength of a surrounding static magnetic field.

The permeable core is driven in and out of saturation by a signal applied to a drive coil (drive winding) on the same core. As the core is driven into and out of saturation, the magnetic flux lines from the external magnetic field being measured are also drawn into and out of the core. These varying flux lines induce positive

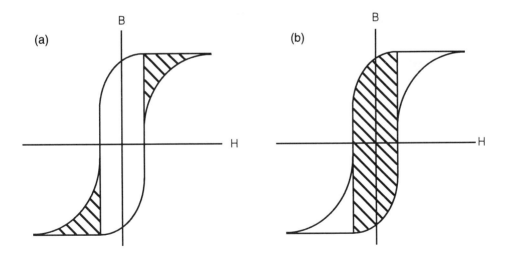

**Figure 3.7** Magnetization (hysteresis) curves for high-permeability materials: (a) in saturation and (b) out of saturation.

and negative electrical current spikes in a sense coil (sense winding). Whenever the flux lines are forced out of the core, positive spikes are generated, and vice versa. The amplitude of these spikes is directly proportional to the intensity of the flux vector. The flux vector, whose direction is determined by the polarity of the pulse, runs parallel to the axis of the sense coil. The greater the difference between the saturated and unsaturated states, the more sensitive the compass will be.

Two sensing coils at right angles to each other have been developed to measure all 360 degrees of the Earth's magnetic field. These two orthogonal pickup coils can be configured in a symmetrical fashion around a common core. A toroidal drive winding is used to excite the permeable core into and out of saturation. More complete flux linkage throughout the core can be achieved using a ring geometry, which in turn requires a smaller drive excitation power and reduces interference in the sense coil(s). For these reasons, as well as ease of manufacture, the many low-cost fluxgate compasses use toroidal ring-core configuration. A sketch of such a compass is shown in Figure 3.8.

The integrated dc output voltages $V_x$ and $V_y$ from the orthogonal coils vary as sine and cosine functions of $\theta$. The instantaneous value of $\theta$ can be easily derived by performing two successive A/D conversions on these voltages, and taking the arctangent of the quotient:

$$\theta = \arctan\frac{V_y}{V_x}$$

where $\theta$ is the angle of the sensor unit relative to the Earth's magnetic field.

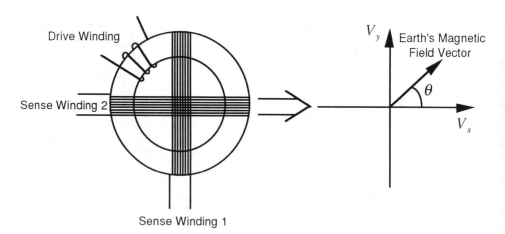

**Figure 3.8** Fluxgate compass.

When the compass is installed in a vehicle it measures the orientation of the vehicle relative to the magnetic field. Any vehicle partially made of iron-based materials generates its own magnetic field, and compasses cannot distinguish between the Earth's magnetic field and other magnetic fields present. In short, compasses measure the vector sum of all magnetic fields present (Figure 3.9 shows the situation if we ignore the effects of any magnetic fields other than the two larger vectors.) Therefore, the equation just given cannot be directly used to derive the direction of the vehicle. The Earth's magnetic field vector may be extracted from the compass measurement by subtracting the magnetic field vector of the vehicle from the compass measurement.

Determining the magnetic field vector of the vehicle is simple. As the vehicle moves and changes its orientation relative to the Earth's magnetic field vector, the vehicle's own magnetic field vector moves with the vehicle. That means that regardless of vehicle orientation, the compass will "see" the sum of the vehicle vector (which remains constant) and a changing Earth vector. If the vehicle is moved through a complete 360 degrees closed loop, the compass output will trace a circle. The coordinates of the center of the circle will describe the vehicle magnetic field vector.

Several methods are available for finding the center of the circle. The simplest one is to drive the vehicle along a closed-loop path and record the maximum and minimum values measured by the $x$ and $y$ coils. Figure 3.10 shows the values generated from an actual trial. The two compass coils might not have identical gain, so the individual gains should be calculated using the following equations:

$$K_x = \frac{X_{max} - X_{min}}{2}$$

$$K_y = \frac{Y_{max} - Y_{min}}{2}$$

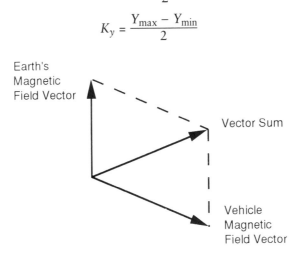

**Figure 3.9** Measurement of vector sum.

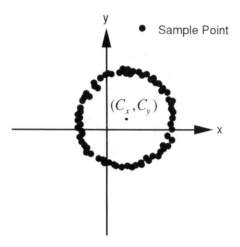

**Figure 3.10** Constructing a magnetization circle to determine the magnetic vector of the vehicle.

The coordinates of the center of the circle are midway between the maximum and minimum points and can be derived from the following equations:

$$C_x = K_x + X_{min}$$
$$C_y = K_y + Y_{min}$$

The direction of the vehicle may be computed using:

$$\theta = \arctan\frac{K_y(V_y - C_y)}{K_x(V_x - C_x)}$$

This method is sensitive to magnetic anomalies or noise. For example, if a truck is positioned along the calibration path such that the magnetic field of the truck affects the maximum or minimum values, the calculated coordinates determined for the center of the circle will contain these errors. Therefore, calibration should be performed in a magnetically quiet area.

In general, a fluxgate compass is very sensitive to magnetic anomalies. Figure 3.11 shows direction change data collected in a very noisy magnetic environment for a vehicle traveling in a straight line. For comparison, we also show the direction change data collected simultaneously using a differential odometer. The driver has tried to follow a straight line. However, it is very difficult to control the steering wheel to force the wheels to follow a perfectly straight line. This fact is reflected in the data collected by the odometer. For the compass data, these direction

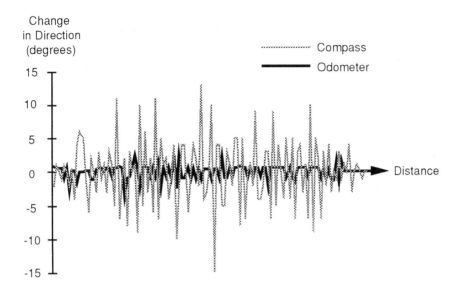

**Figure 3.11** Effect of noisy magnetic environment on direction measurements.

errors result in positioning errors. Therefore, minimization of measurement errors is crucial for any location and navigation system using a fluxgate compass.

Fluxgate compass measurements generally contain two types of errors: short-term magnetic anomalies and long-term magnetic anomalies. Deviations caused by nearby power lines, big trucks, steel structures (such as freeway underpasses and tunnels), reinforced concrete buildings, and bridges are typical examples of short-term magnetic anomalies. Long-term magnetic anomalies result from inaccuracies in calibration, electrical/magnetic noise, or magnetization of the vehicle body.

Filters such as the low-pass filter or complementary filter discussed in Section 3.5 and other valid methods can be used to detect and eliminate short-term magnetic anomalies. These methods include checking the heading change reported by the compass against the physical operating envelope of the vehicle and comparing the measured vector magnitude against the expected range of the Earth's magnetic field. When the external field is strong, it produces a measurement vector that falls far from the curve defined by the circle (Figure 3.10). If we use only samples that fall close to the magnetization circle curve to determine the heading, this source of errors will be eliminated. Relative sensors can also be used to detect blunders (anomalies). When a digital map is available, road segment direction is also a very good reference for detection and correction of compass blunders. Relative sensor data and segment direction are not likely to change dramatically. These blunders are not difficult to detect using if-then rules with proper direction change thresholds relative to the abnormal compass changes associated with a blunder [16].

Errors in calibration are deterministic, and can be eliminated by introducing correction offsets. It is important to realize that calibration errors can change over time. For example, when the vehicle is run through a car wash, the automatic car wash brushes can cause the magnetic field of the vehicle to change for few days so that the center of the circle is shifted (Figure 3.12), which will in turn severely affect the performance of the positioning module. In addition to large moving or stationary steel objects, electronic devices in the vehicle may also cause errors. For instance, when a rear window defroster is in operation, the large current generated may affect a compass mounted nearby. Therefore, care must be taken when deciding where the compass should be installed in the vehicle.

Errors due to certain long-term magnetic anomalies, such as the magnetization of the vehicle body, are difficult to detect. A heading reference that does not use magnetic effects (e.g., GPS) can be used to detect and remove such errors. Electrical or magnetic noise can be averaged over distance and time, and the white noise variance can be estimated *a priori* or estimated in real time when an external heading change reference is available (e.g., differential odometer). As discussed before, the actual magnetic field direction measured by a compass is determined by the combined magnetic field from the Earth and the vehicle itself. If the vehicle is not magnetized, the normalized coil voltages form a circle with the center at the origin. If it is magnetized, the center of the magnetization circle will drop away from the origin (Figure 3.10). A shift curve to 10% of the circle radius will cause a maximum heading error of 6 degrees. The heading error due to the shift in the center of the circle is a function of the actual center and the true heading.

Two methods can be used to correct for the error due to a shift in the center of the magnetization circle. One method is to continuously search for new maximum

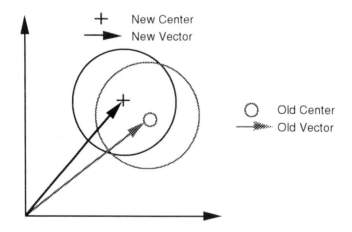

**Figure 3.12** Shift of the vehicle magnetic field center.

and minimum points $V_x$ and $V_y$ to reconstruct the circle center dynamically. Another method is to introduce a correction table. The correction table contains the correction information for any heading. For instance, in Figure 3.13 the compass has an error of −4 degrees at 100 degrees, thus the table will contain a +4-degree correction in the cell for a 100-degree heading. Adding the correction to the compass measurement will eliminate the error. Independent sources of heading information (i.e., GPS or a digital map) can be used for determining the correction and generating the table. These heading sources are often more accurate than the compass but are not available on a continuous basis.

When the vehicle is traveling over a hilly road, the compass plane will not be parallel to the plane of the Earth's surface. The compass measures only the projection of the vector components. This type of error can be detected and removed via a check on the magnitude of the magnetic field.

### 3.4.2 Global Positioning System

The GPS is a satellite-based radio navigation system. It provides a practical and affordable means of determining position, velocity, and time around the globe. GPS was designed and paid for by the U.S. Department of Defense (DOD). Civilian access is guaranteed through an agreement between the DOD and the Department of Transportation (DOT).

GPS consists of three main parts: the space segment (satellites), the user segment (receivers), and the control segment (management and control). From the location

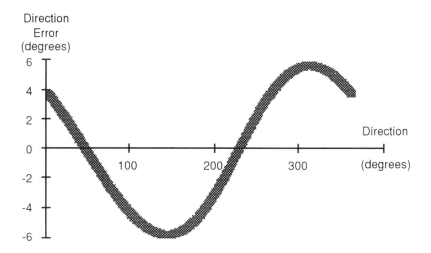

**Figure 3.13** Errors due to the shift of the vehicle magnetic field center.

and navigation point of view, only the first two segments are of interest. For more detailed descriptions of all the segments, as well as various theoretical and practical aspects of GPS, readers may consult [17–19].

The observation of at least four satellites simultaneously will permit determination of the 3D coordinates of the receiver (Figure 3.14) and the time offset between the receiver and GPS time. The 3D coordinates can be computed in an ECEF Cartesian coordinate frame or in a geodetic coordinate system such as WGS 84. Using transformation equations, positions can be computed in other geodetic frames or mapping planes (see Section 2.3).

The GPS constellation consists of 24 satellites arranged in six orbital planes with 4 satellites per orbital plane as shown in Figure 3.15. It is designed to provide worldwide coverage 24 hours per day. The characteristics of GPS are summarized in Table 3.3 (some items in Table 3.3 are from [20]). The design life span of each GPS satellite is 7.5 years. The on-board time and frequency standards are the most likely functions to fail, so each satellite carries four atomic clocks on board.

In Table 3.3, SPS stands for Standard Positioning Service, PPS stands for Precise Positioning Service, and 2dRMS stands for two times the root mean square distance error (twice the standard deviation). The term 100m 2dRMS means that 95% of the time the horizontal positions will be within a circle of a 100m radius. Both the C/A code and P code are pseudorandom noise (PRN) ranging codes. PRN is a sequence of digital 1's and 0's that appear to be randomly distributed like noise, but which can be exactly reproduced. The navigation message broadcast includes satellite orbital position data and satellite health status.

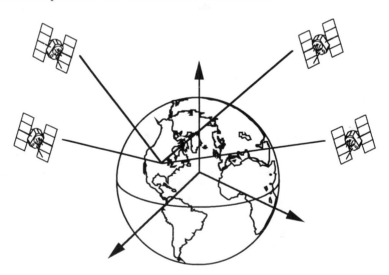

**Figure 3.14** ECEF Cartesian coordinate system for GPS.

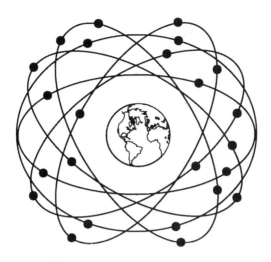

**Figure 3.15** Satellite constellation (space segment).

**Table 3.3**
GPS Characteristics

| *Item* | *Characteristic* |
|---|---|
| Satellites | 24 satellites broadcast signals autonomously |
| Orbits | Six planes, at 55-degree inclination, each orbital plane includes four satellites at 20,231-km altitude, with a 12-hr period |
| Carrier frequencies | L1: 1575.42 MHz; L2: 1227.60 MHz |
| Digital signals | C/A code (coarse acquisition code): 1.023 MHz P code (precise code): 10.23 MHz Navigation message: 50 bps |
| Position accuracy | SPS: 100m horizontal (2dRMS) and 140m vertical (95%) PPS: 21m horizontal (2dRMS) and 29m vertical (95%) |
| Velocity accuracy | SPS: 0.5–2 m/s observed PPS: 0.2 m/s |
| Time accuracy | SPS: 340 ns (95%) PPS: 200 ns (95%) |

The SPS is based on the C/A code and is subject to selective availability (SA). SA is an accuracy degradation scheme to reduce the accuracy available to civilian users to a level within the national security requirements of the United States. It is achieved by a combination of orbital parameter degradation and satellite clock parameter dithering. This results in an intentional error in the navigation message regarding the location (ephemeris) of the satellites and the intentional error in the timing source (carrier frequency) that creates an uncertainty in the Doppler velocity measurements. For civilian users, only SPS is available, therefore, SA degrades the performance of civilian GPS receivers. A simple way to detect whether SA is on or off is to record stationary GPS position and velocity data over a period of time. If SA is off, the 2D position error trace will fall within a circle with a 25m radius (versus 100 m), and the peak velocity error will be about 0.25 m/s (versus 1 m/s). The satellites also broadcast a parameter know as user range accuracy (URA) as part of the navigation message, which can be used as an indication of the status of SA. Recently, the U.S. government has decided to conditionally phase out SA placed on the civilian users (without charging fees) starting in the year 2000, subject to yearly review by the president. This is certainly good news for GPS civilian users.

Differential GPS (DGPS) techniques can improve GPS performance significantly. The technique involves two GPS receivers. One is a master receiver at a reference station with known (surveyed) coordinates and the other is a receiver at a remote location whose coordinates are to be determined (Figure 3.16). The calculated position solution for the master receiver can be compared to the known coordinates to generate a differential correction for each satellite. In one such commonly used technique, the correction information can then be transmitted to the remote receiver for calculating the corrected local position. This effectively reduces the position error

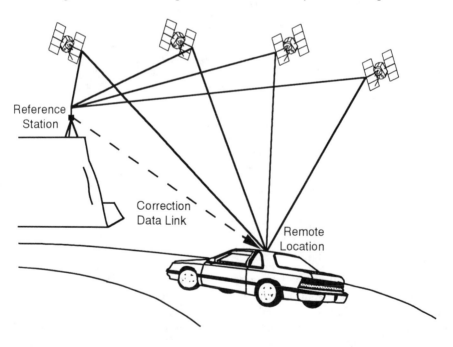

**Figure 3.16** Differential GPS.

for commercial systems to under 15m even when SA is in effect. DGPS technology assumes that the range errors for the remote receiver are the same as for the master receiver. It is easy to see that this assumption may not be true all the time, especially when the two receivers are far apart. An accuracy level of 5m can commonly be achieved for receiver separations of up to 50 km when low-noise receivers are used.

Typical DGPS characteristics are listed in Table 3.4. Comparison with Table 3.3 indicates that DGPS is much better than SPS and even (surprisingly) better than PPS. A more detailed description of DGPS techniques follows the presentation of the basic positioning concepts.

We now briefly discuss the measurement principles used by the GPS receiver. As shown in Figure 3.15, a constellation of satellites is maintained in Earth orbit. Radio receivers at or near the surface of the planet are used to decode the transmitted satellite signals. These receivers then calculate position, velocity, and time from the signals. This is a one-way broadcast system, so receivers do not need to transmit any signals back to the satellites. Since it is a passive system, GPS can support an unlimited number of users, much like a television broadcast system.

Position measurement is based on the principle of time of arrival (TOA) ranging. The time interval taken for a signal transmitted from an emitter (in our case satellite) at a known location to reach a receiver is multiplied by the speed of the signal to obtain the emitter-to-receiver range (distance). Multiple signals received by a receiver from multiple emitters at known locations are used to determine its location. Because of clock offset between satellite and receiver, propagation delays, and other errors, it is impossible to measure the actual range, so a pseudorange is measured. The clock offset is the constant difference in the time reading between the satellite clock and receiver clock. To determine the position of the receiver, the receiver needs to know the pseudoranges to the satellites being tracked and the positions of those satellites. Pseudoranges are obtained by multiplying the apparent signal propagation time by the speed of light. The signal propagation time is determined by measuring the time offset required to match the received satellite code to an internally generated replica of the code. This is known as "correlation." As in all surveying situations, the number of independent observations required depends on how many unknowns need to be solved.

**Table 3.4**
DGPS Characteristics

| *Item* | *Real-Time Characteristic* |
| --- | --- |
| Receiver separation | <50 km |
| Data link update rate | 5–10 sec |
| Position accuracy | 15m (2–5m for low-noise receivers) |
| Velocity accuracy | 0.1 m/s |
| Time accuracy | 100 ns |

In the 2D case, at least two satellites are required to derive the receiver position. If the positions of the satellites are known and the pseudorange from the receiver to each satellite can be measured, the receiver position is constrained to be one of two points of intersection between circles having radii equal to the pseudoranges (Figure 3.17).

The unknown receiver position $(x, y)$ can be calculated using the following equations. Note that these equations use a 2D Cartesian coordinate system for the reference coordinate frame.

$$p_1 = \sqrt{(x - x_1)^2 + (y - y_1)^2}$$

$$p_2 = \sqrt{(x - x_2)^2 + (y - y_2)^2}$$

where $(x_1, y_1)$ and $(x_2, y_2)$ are the known coordinates of the satellites, and $p_1$ and $p_2$ are the measured pseudoranges. After solving these two simultaneous equations, a rough initial estimate (or a third satellite) can be used to determine which of the two solutions should be used.

For the 3D case, three satellites are required (Figure 3.18). Now instead of circles, the pseudoranges yield spheres of position. Clearly, a unique receiver position $(x, y, z)$ of the receiver can be determined by solving the following equations:

$$p_1 = \sqrt{(x - x_1)^2 + (y - y_1)^2 + (z - z_1)^2}$$

$$p_2 = \sqrt{(x - x_2)^2 + (y - y_2)^2 + (z - z_2)^2}$$

$$p_3 = \sqrt{(x - x_3)^2 + (y - y_3)^2 + (z - z_3)^2}$$

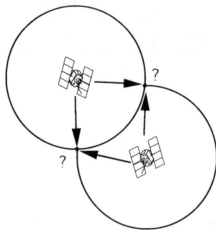

**Figure 3.17** Two-pseudorange intersection.

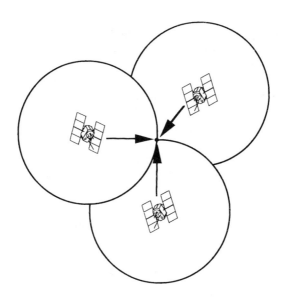

**Figure 3.18** Three-pseudorange intersection.

where $(x_1, y_1, z_1)$, $(x_2, y_2, z_2)$ and $(x_3, y_3, z_3)$ are the known positions of the satellites, and $p_1$, $p_2$ and $p_3$ are the measured pseudoranges.

The receiver clock used to measure the signal propagation times is not synchronized to GPS time. Instead, the clock offset between receiver time and GPS time must be determined. This parameter can be calculated by adding a fourth satellite. By design, all of the satellite clocks are synchronized using very precise atomic clocks. If the receiver clock could be precisely synchronized with the satellite clocks, time measurement would be simple. However, it is economically impractical for receivers to use atomic clocks, so instead, inexpensive crystal oscillators are used. These clocks introduce a time offset (clock bias) between the receiver and the GPS clocks so we must consider their effects in the computation. The receiver clock bias is the time offset of the receiver, and it is the same for each satellite. Thus both the receiver position and clock offset can be derived from the following equations:

$$p_1 = \sqrt{(x - x_1)^2 + (y - y_1)^2 + (z - z_1)^2} + c(dT_1 - dt)$$

$$p_2 = \sqrt{(x - x_2)^2 + (y - y_2)^2 + (z - z_2)^2} + c(dT_2 - dt)$$

$$p_3 = \sqrt{(x - x_3)^2 + (y - y_3)^2 + (z - z_3)^2} + c(dT_3 - dt)$$

$$p_4 = \sqrt{(x - x_4)^2 + (y - y_4)^2 + (z - z_4)^2} + c(dT_4 - dt)$$

where $(x_1, y_1, z_1)$, $(x_2, y_2, z_2)$, $(x_3, y_3, z_3)$, and $(x_4, y_4, z_4)$ are the known satellite positions, $p_1$, $p_2$, $p_3$, and $p_4$ are measured pseudoranges, $c$ is the speed of light, $dT_1$, $dT_2$, $dT_3$, $dT_4$ are the known satellite clock bias terms from GPS time, and $dt$ is the unknown receiver clock offset from GPS time. The satellite clock bias terms are calculated by the receiver from the broadcast navigation message. In the preceding equations, several error terms have been left out for simplicity. For instance, the range errors due to ionospheric delay and tropospheric delay can both be estimated using atmospheric models. However, receiver noise, multipath propagation error, satellite orbit errors, and SA effects remain. The square-root term represents the geometric range between the receiver and satellite, and all other terms contribute to the measurement being a pseudorange.

The preceding equations can be solved by standard techniques such as the Newton-Raphson method. If more than four satellites are used for a solution, a least squares algorithm (Appendix D) or a multiple-state Kalman filter algorithm (Appendix E) can be used. The algorithm used and how it is used have a large impact on the accuracy of position determination and the smoothness of the receiver trajectory.

Let us use four satellites as an example to explain briefly how to solve these equations and to derive measurements of the satellite geometry with respect to the receiver. After linearization, the above equations can be written in a vector-matrix form as

$$\mathbf{Y} = A\mathbf{X}$$

where $\mathbf{Y}$ is the known vector of measurements and $\mathbf{X}$ is the unknown vector. Linear algebra can be used to solve for the unknown vector $\mathbf{X}$. A by-product of the computation is the cofactor matrix

$$P = (A^T A)^{-1}$$

where $T$ stands for the matrix transpose operation and $-1$ stands for matrix inversion. For a 3D GPS solution, the cofactor matrix is a $4 \times 4$ matrix where three components are contributed by the receiver position $(x, y, z)$ and one component by the receiver clock. The matrix can be written as

$$P = \begin{bmatrix} \sigma_x^2 & \sigma_{xy} & \sigma_{xz} & \sigma_{xt} \\ \sigma_{yx} & \sigma_y^2 & \sigma_{yz} & \sigma_{yt} \\ \sigma_{zx} & \sigma_{zy} & \sigma_z^2 & \sigma_{zt} \\ \sigma_{tx} & \sigma_{ty} & \sigma_{tz} & \sigma_t^2 \end{bmatrix}$$

To use a GPS receiver's output properly, users must understand various potential GPS errors such as the uncertainty in satellite positions and the effects of propagation conditions. In addition, the geometrical position of the receiver with respect to the satellites being tracked plays a considerable role in GPS receiver measurement errors. Various dilution of precision (DOP) measures are used to characterize the geometric contribution to the receiver error. The calculation of these DOPs is based on the square root of the sum of the variances of the errors in the estimated position along the three coordinate axes, and of the estimated time as shown below. We usually consider the GPS geometry good if the GDOP (geometrical DOP) is less than 3 or PDOP (position DOP) is less than 12.

$$\text{GDOP} = \sqrt{\sigma_x^2 + \sigma_y^2 + \sigma_z^2 + \sigma_t^2}$$

$$\text{PDOP} = \sqrt{\sigma_x^2 + \sigma_y^2 + \sigma_z^2}$$

Velocity measurements are based on the Doppler frequency shift principle. The Doppler shift in the frequency of each satellite is a direct measure of the relative velocity of the receiver and the satellite along the line between them. Each satellite has a very high velocity relative to a stationary receiver, both because of the orbital motion of the satellite and because of the motion of the receiver associated with the Earth's rotation. The velocity solution can be calculated by differentiating the four-dimensional navigation solution (pseudorange equations) introduced earlier with respect to time.

To have a general idea of the composition of a GPS receiver, Figure 3.19 depicts a block diagram of a Motorola Oncore GPS receiver, where ASIC stands for application specific integrated circuit and EEPROM stands for electronically erasable programmable ROM. See [21] for more information about the receiver design and some practical receiver examples.

Readers who are interested in evaluation and selection of GPS receivers for specific applications should consult [22] and [23]. For location and navigation applications, GPS receiver performance under heavy foliage and in urban canyon areas is particularly important. Overhead foliage may cause significant attenuation of satellite signals, and high-rise buildings in urban canyons may block and reflect satellite signals. Therefore, a receiver with a position filter, fast reacquisition time, and parallel channel tracking (versus multiplexing channel) capabilities is much better than one without such features (multiplexing is the practice of rapidly sequencing any or all of the available channels among the visible satellites). For performance test reports under these circumstances, see [24,25]. Another feature that enhances GPS receiver performance is receiver autonomous integrity monitoring (RAIM). It provides detection, isolation, and removal of faulty satellite signals. Finally, research indicates that an eight-channel low-cost receiver provides an excellent balance of cost and performance needed to track satellite signals reliably.

As mentioned earlier in this section, when determining the receiver position, several errors are associated with the pseudorange measurement. Table 3.5

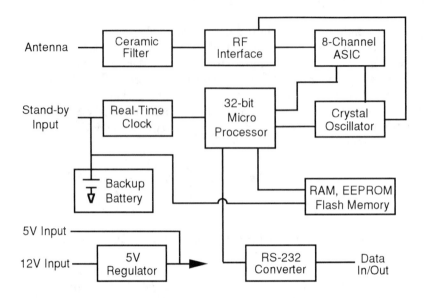

**Figure 3.19** Motorola Oncore GPS receiver diagram.

Table 3.5
GPS Receiver Errors

| Error | Typical GPS (m) | Worst-Case GPS (m) | Typical DGPS (m) |
|---|---|---|---|
| Satellite clock | 2–3 | 25 | <0.5 |
| Satellite orbit | 1–2 | 5 | <0.5 |
| SA | 30–50 | 100 | <0.5 |
| Ephemeris prediction | 3–5 | 15 | <0.5 |
| Ionospheric delay | 10–15 | 100 | <0.5 |
| Tropospheric delay | 3–5 | 30 | <0.5 |
| Multipath | 10 | 300 | 10 |
| Receiver noise | 5 | 15 | 5 |
| Total effect | 100 (95%) | 300 (99.99%) | 15 (95%) |

summarizes the typical effects of these errors on the position output by the receiver. From the table, we see that there are other sources of error in addition to SA. The first six errors in Table 3.5 are virtually zero for typical DGPS (receiver separation less than 50 km) because they are highly correlated.

Other techniques can also be used when designing the receiver algorithm to reduce the errors further. For civilian receivers, two methods can be used to calculate receiver position at first lock-on to satellites. One is based on the code phase (1.023 MHz, wavelength ≈ 293,000 m). The other is based on the carrier phase

(1575.42 MHz, wavelength $\approx$ 0.19 m). Both the code and the carrier phases are transmitted by the satellite as shown in Table 3.3. The receiver generates its own code or carrier to correlate with the one received from the satellite in order to decode the navigation message. Both methods have associated range ambiguities because the receiver needs to decide the number of cycles transmitted from the satellite in order for the receiver to initially lock on. For the code phase tracking, the ambiguity is easily resolved.

Sophisticated algorithms are required to resolve the carrier phase tracking ambiguity. A combination of these two observations, as is done in carrier-aided code tracking (carrier-smoothed code) produces better results. For instance, typical Motorola Oncore DGPS can reduce both the multipath and receiver noise errors down to 1 to 2m. One study shows that by postprocessing the raw Oncore output data, DGPS accuracy can be made as good as 2 cm [26]. The postprocessing uses carrier phase differential processing and ambiguity resolution techniques.

GPS receivers can be used for time synchronization and frequency stabilization. Recall that the satellites have very expensive on-board atomic clocks with picosecond accuracy ($10^{-12}$) compared with the nanosecond accuracy ($10^{-9}$) of low-cost GPS receivers. Furthermore, the clock time-keeping ability of the satellites is better than other technologies because all satellite clocks are time synchronized by the control segment to GPS time. The time difference between any two GPS receivers is typically within 60 ns. Therefore, it is very economical to use low-cost GPS receivers to synchronize time to either GPS time or universal time coordinated (UTC) time. Examples of such applications are to synchronize time between adjacent cellular telephone cells to smooth the transition of a caller from one cell to another, and to synchronize time among three radio-frequency reference sites to locate a mobile user (Sections 9.2.2 and 10.2). Because GPS receivers can generate steady pulses (one pulse per second or 1 Hz), these pulses can be used for comparison with oscillator frequencies. Using a correction circuit, the frequency of the oscillator can be stabilized very precisely.

GPS receivers have many different applications. In geodetic surveying, a reference station receiver with few mobile receivers at selected surveying locations can collect raw data in the field, and the resulting data can be used to calculate the distances between the receiver locations in the office. This is called a *static postprocessing survey*. A reference station receiver with one mobile receiver can collect data while the mobile receiver is moved to different surveying locations in the field, and the distances are calculated in the office. This is known as a *semi-kinematic* or *rapid static survey*. A reference station receiver or a mobile receiver can broadcast its raw measurements to the other receiver, so that carrier phase differential processing can be done in real time. This method is called a *real-time kinematic (RTK) survey*.

DGPS has different implementations. For instance, if both the satellite positions and the receiver positions are known, the range correction vector can be derived by subtracting the known range vector from the measured range vector. The correction vector can then be transmitted from the reference station to the mobile receiver. The data transmission rates for the corrections are determined by the rates of variation in the errors. For autonomous vehicle location and navigation, one-way communication of corrections is sufficient. For vehicle tracking, two-way communication may be required if the system needs both the user in the vehicle and a central dispatcher to have corrected positions. VHF (very high frequency) radio, two-way radio, FM subcarrier radio, pager signals, cellular telephone, or satellite-based communications, etc., can be used for this communications link. We discuss RF communication in more detail in Chapter 8.

One cost-effective technique for vehicle tracking is inverse DGPS. In this situation, the vehicle does not have the corrected positions; only the reference station does. This technique requires the vehicle to transmit satellite numbers and ephemeris data for the satellites it is tracking to the reference station along with the vehicle position. The software in the reference station uses these data and other data collected by the reference station using the same satellites to compute a position fix. It then derives the correction vector for that vehicle and applies it. If the system needs to track many vehicles, the reference station needs a very powerful processor or processors because of the computationally intensive nature of the task.

Wide-area DGPS (WADGPS) is another cost-effective method of bringing differential techniques to many users. Recall that DGPS needs to have a receiver separation of less than 50 km to provide very good accuracy. In other words, the main disadvantage of DGPS is the high correlation between the reference-station-to-user distance and the positioning accuracy achieved. To cover a wide area, a substantial increase in the number of reference stations is required. In contrast, WADGPS requires very few reference stations. For example, 10 sparsely located reference stations may be enough to cover the continental United States. Using the GPS receiver data collected by these 10 stations, a national map can be calculated for the entire country. Whenever any local user needs the service, this national map can be used to derive corrections for a virtual reference station in that particular area. Since this virtual reference station is nearly equivalent to a real reference station in the same area, similar position accuracy can be achieved. Research has also been very active in applying similar techniques to determine the exact satellite orbits. If real-time computation can be achieved, this *precise orbits* method should be able to improve the positional accuracy even more. Currently, precise orbits and clocks are available for postmission use only.

In addition to the Global Positioning System developed by the United States, other satellite systems have been developed that could be used for location and navigation, such as GLONASS, which was developed by the former USSR. NAVSAT has been developed by the European Space Agency (ESA), LOCSTAR is being promoted by the French space organization (CNES), and STARFIX has been developed by John E. Chance and Associates of Louisiana for the Gulf of Mexico and continental United States. More information about these systems can be found in [21,27] and the references cited therein. We discuss other satellite communication technologies in Section 8.3.8; most of the satellite systems in operation or in planning also offer position determination as a feature. In general, the positional accuracy provided by these technologies is worse than that of GPS.

## 3.5 SENSOR FUSION

From the preceding discussion, it is clear that no single sensor can provide completely accurate vehicle position information. Therefore, multisensor integration is required in order to provide the on-vehicle system with complementary, sometimes redundant information for its location and navigation task. Many fusion technologies have been developed to fuse the complementary information from different sources into one representation format. The information to be combined may come from multiple sensors during a single period of time or from a single sensor over an extended period of time. Integrated multisensor systems have the potential to provide high levels of accuracy and fault tolerance.

Multisensor integration and fusion provide a system with additional benefits. These may include robust operational performance, extended spatial coverage, extended temporal coverage, an increased degree of confidence, improved detection performance, enhanced spatial resolution, improved reliability of system operation, increased dimensionality, full utilization of resources, and reduced ambiguity. However, we must keep in mind that although there are many advantages in sensor fusion, most fusion methods make certain assumptions, either explicitly or implicitly. If the assumed sensor model does not adequately describe the data from the real sensor, a perfect sensor modeling and estimation theory may not produce the expected result.

### 3.5.1 Simple Filters

Because of error propagation, many different error filtration methods are used. One such method is the digital filter. A low-pass filter can be used to reduce the disturbing effects of other magnetic fields on the compass. A high-pass filter can be used to eliminate errors due to odometer drift. The output of these two filters can then be integrated to obtain the output from the positioning module. From the basic deterministic methods of linear system analysis, a low-pass filter can be written in the form of a transfer function:

$$G(s) = \frac{1}{1 + \tau s}$$

where $\tau$ is the time constant of the low-pass filter. If a hardware filter is used, $\tau$ could be the product of a resistance and a capacitance. By substituting $s$ with $1/s$, we can derive the transfer function for the high-pass filter:

$$G(s) = \frac{s}{s + \tau}$$

Both filters can be implemented in either hardware or software. One such implementation is discussed in [28].

If the sensor outputs come from two independent noisy measurements of the same signal, a complementary filter can be used to reduce the noise without distorting the signal [29,30]. The filter is shown in Figure 3.20, where $x(t)$ is the measured signal, $n_1(t)$ and $n_2(t)$ are the noises on each signal, $y(t)$ is the output of the complementary filter, and $G(s)$ is the transfer function, which represents the dynamic characteristics of the underlying sensor system. This technique is very effective when the spectral content of the noise is mainly at low and high frequencies, respectively.

The purpose of this filter is to perform noise cancellation and to obtain an output $y(t)$ as close as possible to the original signal $x(t)$. Similarly, from the basic deterministic methods of linear system analysis and the block diagram in Figure 3.20, it is not difficult to see that the Laplace transform of the filter output is

$$Y(s) = X(s) + N_1(s)[1 - G(s)] + N_2(s)G(s)$$

Clearly, the signal term $X(s)$ is not affected by $G(s)$. Therefore, we can select a $G(s)$ that attenuates both noise components.

### 3.5.2 Kalman Filters

Kalman filtering uses the statistical characteristics of a measurement model to recursively estimate fused data. These data are optimal in a statistical sense. If a system can be described using a linear model and both the system and sensor errors can be modeled using Gaussian white noise, the Kalman filter will provide statistically optimal estimates for the fused data. The filter can be used without storing any historical data because of its recursive nature. It can also be used to blend location determined by the positioning and the map-matching modules (Chapter 4) to yield a best estimate for the position.

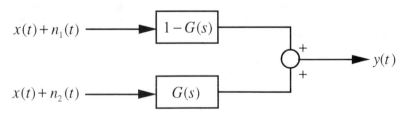

**Figure 3.20** Complementary filter.

A Kalman filter can be used to fuse measurements from multiple sensors and provide both an estimate of the current state of a system and a prediction of the future state of the system. Two approaches can be taken: a centralized (standard) Kalman filter or a decentralized Kalman filter. A centralized filter takes all the sensor input and outputs a complete solution while a decentralized filter uses local filters for each sensor and a master filter to fuse all the sensors (Figure 3.21). When applied to multisensor systems and/or systems with embedded local filters, research results show that a centralized filter may result in heavy computation loads, poor fault tolerance, and an inability to correctly process prefiltered data in a cascaded filter structure. During the last several years, many decentralized Kalman filters have been introduced in an effort to improve fault tolerance (see [31] and its references).

Many good specialized textbooks have been published that discuss the underlying principles and implementation of Kalman filters [29,32,33]. The recursion equations for discrete Kalman filters are given in Appendix E. As mentioned earlier, the fusion method generally makes certain assumptions. This is certainly the case for a Kalman filter. It assumes that all measurement errors not modeled are uncorrelated and that the variations in these errors are uncorrelated (white noise). Therefore, if these assumptions are not valid, a theoretically perfect filter algorithm might lead to unacceptable results. Since the discrete Kalman filter is recursive, round-off error may lead to problems as the number of steps increases. Sometimes there may be one or more unobservable state variables. All of these situations may cause the system to diverge.

Many other Kalman filters can be used for sensor fusion. In some applications, a linear Kalman filter can be used, provided the nominal trajectory can be linearized in the state space of the system (i.e., the system model is approximately linear) and this trajectory does not depend on the results of the measurement. An extended Kalman filter can be used if the estimated trajectory can be linearized under the assumption that this trajectory can be estimated from the measurements [12]. If the estimated trajectory of the extended Kalman filter is worse than the normal one, this could lead to further error in the trajectory, which causes further errors in the estimates, and so on, leading to the eventual divergence of the filter. Therefore, caution must be used, especially in the situation where the initial uncertainty and measurement errors are both large. An adaptive Kalman filter can be used if all the parameters of the filter are not known initially or if the filter parameters change over time. When numerical stability such as round-off error is of special concern for possible divergence of the filter, a U-D covariance factorization filter can be used [34]. U is an upper triangular matrix (ones along the major diagonal, nontrivial elements in the upper triangular part, and zeros in the lower triangular part) and D is a diagonal matrix. In practice, many systems have no divergence problems, because they are completely observable and there is adequate process noise feeding into all system states.

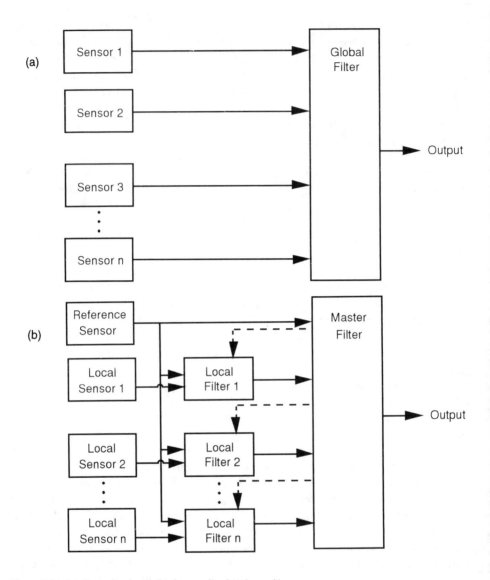

**Figure 3.21** (a) Centralized and (b) decentralized Kalman filters.

### 3.5.3 Other Fusion Methods

Many other techniques have been devised that can be used for sensor fusion [7,35]. Although the names of some of these methods may not be familiar, we introduce them here for interested readers to explore further. Many techniques not mentioned here are often used for pixel processing and image recognition and could be very useful for enhancing drivers' vision or even allowing automatic driving. Some techniques have already been proposed and used in practice.

Fuzzy logic can be used to represent uncertainty in the inference (or fusion) process to make a crisp decision. A distributed blackboard architecture can be used to ease the communication between various sensors in the fusion process. Weighted averaging can be employed to average redundant information. Statistical decision theory can be used to model sensor noise as probability distributions for removal of outliers. Bayesian estimation (Bayesian inference) can be applied to combine sensor information based on the rules of probability theory. Dempster-Shafer evidential reasoning can be used to extend the Bayesian estimation. Production rules can be used to symbolically represent the relation between sensory information and an attribute that can be inferred from the information as part of an expert system, etc. A neural network can be trained for fusing sensors without knowing their characteristics. Neural networks consist of many layered nodes (neurons) that are interconnected. A neural network is a dynamic system and must be trained to rearrange layers and interconnections in order to model the real-world application. One important feature of neural networks is that they simulate a dynamic system without a mathematical model of how the outputs depend on the inputs. Neural networks learn from numerical (and sometimes even linguistic) sample data.

Although various sensor integration and fusion technologies are available, we must keep in mind that any technique used in a vehicle location and navigation system must be efficient enough for real-time derivation of the correct position. Otherwise, the technique is not suitable for actual implementation in an embedded real-time system.

In general, conventional vehicle positioning techniques such as dead reckoning tend to accumulate errors; absolute positioning systems such as GPS are free from error accumulation. However, as discussed, the accuracy and availability of GPS are affected by SA and the vehicle environment. Consequently, the raw vehicle position derived from sensor measurements as described earlier will not be error free. In Chapter 4, we discuss how to use the map-matching module to enhance the vehicle positioning system.

Before turning to the next chapter, can you list the main error sources for each sensor discussed and the compensation methods learned that can eliminate or reduce these errors? Are there any other approaches that can compensate for the errors you listed?

## References

[1] K. Jost, "Sensor Technology Review," *Automotive Engineering*, Vol. 103, No. 9, Sep. 1995, pp. 39–49.
[2] R. Jurgen (Ed.), *Automotive Electronics Handbook*, New York: McGraw-Hill, 1995.

[3]  C. O. Nwagboso (Ed.), *Automotive Sensory Systems*, London/New York: Chapman & Hall, 1993.

[4]  W. B. Ribbens, *Understanding Automotive Electronics*, 4th ed., Carmel, IN: SAMS, 1992.

[5]  J. Fraden, *AIP Handbook of Modern Sensors: Physics, Designs, and Applications*, New York: American Institute of Physics, 1993.

[6]  L. Ristic (Ed.), *Sensor Technology and Devices*, Norwood, MA: Artech House, 1994.

[7]  R. C. Luo and M. G. Kay (Eds.), *Multisensor Integration and Fusion for Intelligent Machines and Systems*, Norwood, NJ: Ablex Publishing Corp., 1995.

[8]  D. A. Foster, "Developments in Wheel Speed Sensing," *ABS Traction Control*, SAE Paper 880325, Society of Automotive Engineers, Aug. 1988, pp. 81–87.

[9]  R. F. Wells, "Non-Contacting Sensors: An Update," *Automotive Engineering*, Vol. 96, No. 11, Nov. 1988.

[10]  S. Oho, H. Kajioka, and T. Sasayama, "Optical Fiber Gyroscope for Automotive Navigation," *IEEE Trans. Vehicular Technol.*, Aug. 1995, pp. 698–705.

[11]  N. Yumoto, "Onboard Equipment," in *Advanced Technology for Road Transport: IVHS and ATT*, Ian Catling (Ed.), Norwood, MA: Artech House, 1994, pp. 338–342.

[12]  B. Barshan and H. F. Durrant-Whyte, "Inertial Navigation Systems for Mobile Robots," *IEEE Trans. Robotics Automation*, Vol. 11, No. 3, June 1995, pp. 328–342.

[13]  J. Miyazaki, R. Higashimae, T. Kuirhara, and S. Ishino, "Vibratory Gyroscope and Various Applications," *Murata Electronics North America*, 1995.

[14]  T. Oikawa, Y. Aoki, and Y. Suzuki, "Development of Vibrational Rate Sensor and Navigation System," *SAE Paper No. 870215*, Society of Automotive Engineers, 1987.

[15]  L. E. Lenz, "A Review of Magnetic Sensors," *Proc. IEEE*, Vol. 78, No. 6, June 1990, pp. 973–989.

[16]  Y. Zhao, L. G. Seymour, and E. M. Kozikaro, "Absolute Heading Sensor Blunder Detection Using Relative Heading Sensor and Road Segment," *Technical Developments*, Motorola, Vol. 29, Nov. 1996, pp. 104–107.

[17]  B. Hofmann-Wellenhof, H. Lichtenegger, and J. Collins, *Global Positioning System: Theory and Practice*, 3rd ed., Berlin: Springer-Verlag, 1994.

[18]  A. Leick, *GPS Satellite Surveying*, New York: John Wiley & Sons, 1995.

[19]  B. W. Parkinson, J. J. Spilker Jr., P. Axelrad, and P. Enge (Eds.), *Global Positioning System: Theory and Applications*, Washington, DC: American Institute of Aeronautics and Astronautics, 1996.

[20]  DoD/DoT, *1994 Federal Radionavigation Plan*, Springfield, VA: National Technical Information Service, Report No. DOT-VNTSC-RSPA-95-1/DOD-4650.5, May 1995.

[21]  B. Forssell, *Radionavigation Systems*, New York: Prentice Hall, 1991.

[22]  J. F. McLellan and J. P. Battie, "Testing and Analysis of OEM GPS Sensor Boards for AVL Applications," *Proc. IEEE Position Location and Navigation Symposium*, Apr. 1994, pp. 512–519.

[23]  Y. Zhao, "Evaluation of GPS Receivers for Automotive Applications," *Proc. Second World Congress on Intelligent Transport Systems*, Nov. 1995, pp. 568–572.

[24]  G. Lachapelle and J. Henriksen, "GPS Under Cover: The Effect of Foliage on Vehicular Navigation," *GPS World*, Mar. 1995, pp. 26–35.

[25]  Motorola, *Navigation Performance Test Report*, Position and Navigation Systems Business, Jan. 1994.

[26]  A. Tabsh, H. Martell, and D. Cosandier, "High Precision Real-Time Positioning Using Low Cost OEM Sensors," *Proc. ION GPS-95*, Sep. 1995, pp. 1433–1441.

[27]  E. Kaplan (Ed.), *Understanding GPS: Principles and Application*, Norwood, MA: Artech House, 1996.

[28]  M. L. G. Thoone, "CARIN, A Car Information and Navigation System," *Philips Tech. Rev.*, Vol. 43, No. 11/12, Dec. 1987, pp. 317–329.

[29]  R. G. Brown and P. Y. C. Huang, *Introduction to Random Signal Analysis and Applied Kalman Filtering*, New York: Wiley, 1992.

[30] W.-W. Kao, "Integration of GPS and Dead-Reckoning Navigation Systems," *Proc. IEEE-IEE Vehicle Navigation & Information Systems Conference (VNIS '91)*, Society of Automotive Engineers, Oct. 1991, pp. 635–643.

[31] H. A. Carlson, "Federated Kalman Filter Simulation Results," *Navigation: J. Institute of Navigation*, Vol. 41, No. 3, Fall 1994, pp. 297–321.

[32] A. Gelb (Ed.), *Applied Optimal Estimation*, Cambridge, MA: MIT Press, 1974.

[33] P. S. Maybeck, *Stochastic Models, Estimation and Control*, New York: Academic Press, 1979.

[34] E. J. Krakiwsky, C. B. Harris, and R. V. C. Wong, "A Kalman Filter for Integrating Dead Reckoning, Map Matching, and GPS Positioning," *Proc. IEEE Position, Location and Navigation Symposium (PLANS)*, IEEE, pp. 39–46, 1988.

[35] W. Waltz and J. Llinas, *Multisensor Data Fusion*, Norwood, MA: Artech House, 1990.

# CHAPTER 4

▼▼▼

# MAP-MATCHING MODULE

## 4.1 INTRODUCTION

The map-matching module plays an important role in vehicle location and navigation systems. It employs a digital map to make the positioning system more reliable and accurate.

To provide drivers with proper maneuvering instructions or to correctly display the vehicle on a map in an error-free fashion, the vehicle location must be precisely known. Therefore, an accurate vehicle location is considered a prerequisite for good system performance. In Chapter 3, we discussed how a vehicle location and navigation system can rely on sensors to determine its position. One such technique is called dead reckoning. Dead reckoning can track the position of a vehicle relative to another position, such as an origin. Typically, vehicle heading and distance traversed are used to determine incremental changes (Section 3.2) in the position of the vehicle relative to the origin. As discussed in Sections 3.2, 3.3, and 3.4, these sensory systems have errors associated with their operation. These errors (especially cumulative errors) detract from the accuracy of the dead-reckoning process. Even with very good sensor calibration and sensor fusion technologies, inaccuracies are often inevitable. As a result of these inaccuracies, the actual vehicle position will not agree with the dead-reckoned vehicle position. These inaccuracies are likely to increase or continue to accumulate with additional distance traversed, thereby increasing the dead-reckoning error. The result is more uncertainty in the true position of the vehicle.

To deal with this uncertainty, map-matching systems have been developed to match the dead-reckoned position (or trajectory) with a position associated with a location (or route) on a map. When the dead-reckoning behavior indicates that vehicle is in a certain position on the map (e.g., on a particular road segment), the vehicle position may be adjusted to some absolute position on the map. This will probably eliminate the cumulative error until the next map-matching step. By doing this process in each successive cycle of the system, a more accurate position can be determined for the system. Figure 4.1 shows a comparison for two consecutive system cycles. To convey the idea, we make the accumulated dead-reckoning errors more visible. In practice, the errors should be smaller for a well-calibrated dead-reckoning system. Although dead reckoning is often used in this chapter for the convenience of the discussion to describe the position sensing and map-matching processes, we should keep in mind that other positioning techniques, such as satellite-based positioning and terrestrial-radio-based positioning, can also be used in place of dead reckoning. Because a map and a matching process are involved, these algorithms are called *map-matching algorithms*.

A good map-matching algorithm can significantly enhance the accuracy of location on a road network. The reason is straightforward. Unlike navigation in air and sea transportation, road transportation vehicles are essentially constrained to a finite network of roads, with only temporary excursions into parking lots, driveways, or other off-road conditions. This makes it possible to use computer algorithms to correlate the trajectory of the vehicle with the digital road information discussed in Chapter 2.

A map-matching algorithm can correlate the vehicle position based on various sensors to the road network by comparing the vehicle trajectory with the routes

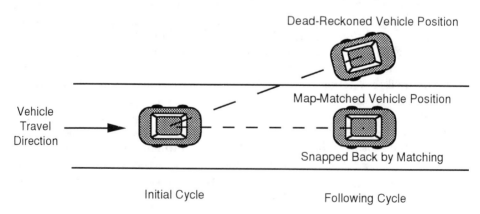

**Figure 4.1** Comparison of the map-matching and dead-reckoning techniques.

present in the digital map database module. Of course, this implies that the system assumes the vehicle to be on a road. In this module, the travel path of the vehicle is constantly compared with the available routes, which consist of a series of connected roads in the road network. Through a pattern recognition and matching process, the most likely location of the vehicle with respect to the map is determined. Because the digital map database contains the coordinates of the roads, the map-matched positions can be used to reset the raw vehicle positions and limit the magnitude of the position errors.

As mentioned, map matching is the process by which the vehicle trajectory (path) is correlated with a digital road map, thereby locating the vehicle relative to this map. This is a software-based technology and makes extensive use of the digital map stored in the system. Clearly, the digital map used for this module must be relatively accurate; otherwise, the system will generate an incorrect position output, which in turn will severely degrade system performance. It is generally considered to be acceptable if the digital map is accurate to within 15m of "ground truth."

## 4.2 CONVENTIONAL MAP MATCHING

The basic idea of conventional map matching is to compare the vehicle's trajectory against known roads close to the previous matched position. The road whose shape most closely resembles the current trajectory and previously matched road is selected as the one on which the vehicle is apparently traveling.

The early conventional map-matching algorithm was invented in the 1970s independently by two U.S. groups and one U.K. group [1]. As time passed, this semi-deterministic method has evolved into a probabilistic method.

### 4.2.1 Semi-Deterministic Algorithms

The semi-deterministic method requires that an initial vehicle location (initial coordinates) and a direction of travel be provided. Conditional tests are intermittently applied to determine whether the vehicle is traveling on the known road network by comparing the dead-reckoned turns to the turns implied by the road segment vectors computed from the map coordinates. Exact correspondence is not possible due to errors in dead-reckoning sensors and in the map coordinates. The module must have a threshold to help limit potential matches. Otherwise, it would be difficult to locate any matching road segment. For this technique to work, the vehicle is generally assumed to be on a road. The accumulated errors in the dead-reckoning distance sensors are corrected whenever the system detects a turn. If the trajectory of the vehicle drifts from the road between turns, the accumulated error of the heading is also corrected. This method is based on the presumption that the vehicle is following a predetermined road. There is considerable uncertainty when the vehicle travels off-road because there is no longer any way for the sensor errors to be corrected. For this reason, contemporary map-matching-based positioning systems tend not to use this method.

### 4.2.2 Probabilistic Algorithms

The probabilistic method requires that the dead-reckoning sensor errors be statistically propagated into the position determination. Elliptical or rectangular confidence regions defined based on these errors and error models are used to outline the boundary within which the true vehicle position may lie. This confidence region is superimposed on the road network. If the region contains only one intersection or road, a match is made and the coordinates are sent to be integrated or further processed as a vehicle position. If more than one intersection or road is in the region, connectivity checks are made to determine the most probable location of the vehicle. The main enhancement of this method over the semi-deterministic method is that it does not require the assumption that the vehicle is always on a road. If the vehicle appears to be off the known road network, this method repeatedly compares the sensor-detected coordinates with those road segments surrounding the off-road area and tries to identify the road segment to which the vehicle appears to have returned. Figure 4.2 contains a flowchart for a typical conventional map-matching algorithm.

Even though the flowchart for this diagram looks simple, many approaches may be used for the actual computation. Within each step, different algorithms can be used. As discussed in Section 3.5, different sensor fusion methods can be used for the first step (or logic box) in Figure 4.2.

As for the second step, many methods are available for calculating the error region associated with a particular positioning system. One way is to derive an error ellipse. From estimation theory, we know that many input and output signals can be modeled as stochastic processes. Variables associated with the true and measured values can be modeled as random variables. Variance-covariance information is propagated through appropriate algorithms to derive the variances and covariances as functions of the original random variables or as functions of parameters estimated from the original observations. These variances and covariances are often used to define error ellipses. To learn more about stochastic processes and estimation theory, readers are encouraged to consult [2–4]. The following simplified discussion should give you a basic idea of how this statistical theory is used in practice.

Suppose that the variance-covariance matrix of the system being modeled is

$$P = \begin{bmatrix} \sigma_x^2 & \sigma_{xy} \\ \sigma_{yx} & \sigma_y^2 \end{bmatrix}$$

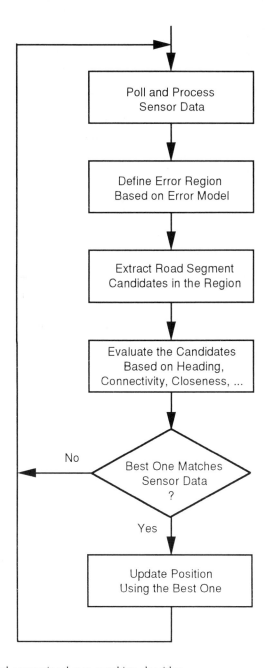

**Figure 4.2** A simplified conventional map-matching algorithm.

where $\sigma_x$ and $\sigma_y$ are the standard deviations of the sensor measurement errors, $\sigma_x^2$ and $\sigma_y^2$ are the variances, and $\sigma_{xy}$ and $\sigma_{yx}$ are the covariances. The error ellipse can be derived as

$$a = \hat{\sigma}_0 \sqrt{\frac{1}{2}\left(\sigma_x^2 + \sigma_y^2 + \sqrt{(\sigma_x^2 - \sigma_y^2)^2 + 4\sigma_{xy}^2}\right)}$$

$$b = \hat{\sigma}_0 \sqrt{\frac{1}{2}\left(\sigma_x^2 + \sigma_y^2 - \sqrt{(\sigma_x^2 - \sigma_y^2)^2 + 4\sigma_{xy}^2}\right)}$$

$$\phi = \frac{\pi}{2} - \frac{1}{2}\arctan\left(\frac{2\sigma_{xy}}{\sigma_x^2 - \sigma_y^2}\right)$$

where $a$ is the length of the semi-major axis of the ellipse, $b$ is the length of the semi-minor axis of the ellipse, $\phi$ is the orientation of the semi-major axis of the ellipse relative to North (Figure 4.3), and $\hat{\sigma}_0$ is the a posteriori variance of unit weight, also known as the expansion factor. Assuming that the distribution of measurement errors is a standard normal distribution, the standard ellipse ($\hat{\sigma}_0 = 1$) corresponds to a 39% confidence region. The dimensions of the ellipse can be scaled to represent various confidence levels. In the 2D case, an expansion factor $\hat{\sigma}_0$ of 2.15 can be used to obtain the 95% confidence level, and 3.03 can be used to obtain a 99% confidence level.

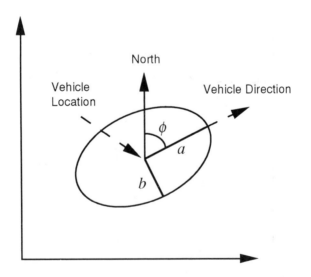

**Figure 4.3**  Error ellipse.

If the data received from a GPS receiver alone are used to define the map-matching error ellipse, we can obtain the variances and covariances directly from the outputs of the GPS receivers. For instance, if an Oncore GPS receiver is used, we can substitute $A_{11}^{-1}$ for $\sigma_x^2$, $A_{22}^{-1}$ for $\sigma_y^2$, and $A_{12}^{-1}$ for $\sigma_{xy}$ and $\sigma_{yx}$, where $A^{-1}$ is the cofactor matrix $P$ discussed in Section 3.4. These parameters are available by invoking the Position/Status/Data Extension Message command [5]. Consulting other GPS receiver manuals may allow the user to identify similar output variables, possibly under different names.

For multisensor-based map-matching systems, different statistical methods have also been developed to define the error region, such as contours of equal probability (CEP) [6], and vehicle location probability area (VLPA) [7]. We discuss the CEP algorithm in further detail later in this chapter. If a Kalman filter (Appendix E) is used for sensor integration and fusion, the task of defining the error region becomes easier. Because the variance-covariance information is available as a by-product of the filter computations, this information can readily be used to derive the error region (ellipse) as noted before.

Note that it is very common to multiply the derived error region by an expansion (confidence) factor. There may be other uncertainties in addition to the sensor errors. Map databases could have errors due to errors in the original sources, measurement errors, processing mistakes, and so on. The initial reference vehicle position might have been slightly different from the actual position. The vehicle could be anywhere within the width of a particular segment or lane. All of these factors contribute to the uncertainty, so expansion of the error region is often appropriate.

To implement the third step in Figure 4.2, it is common to first modify the error region. Since ellipses are inconvenient for locating candidate segments, a rectangle is generally used. To ensure that the error region is large enough to include all potential candidate segments, an expansion factor can once again be used to further enlarge the ellipse to a higher confidence level before the ellipse is clipped to a rectangular region. The Cyrus-Beck clipping algorithm can be used to locate the segments in the region [8]. This algorithm uses the dot products of the inner normals of the rectangular boundaries and vectors from a point on the candidate segment to a point on each boundary to determine whether the segment lies inside or outside of the clipping region.

As for the fourth step in Figure 4.2, we can sort the segments found in the previous step. The segment that most closely matches the given sensor data should be at the top of this list. Appropriateness of the segment heading is a very important criterion. Because of the many uncertainties involved with the sensor and segment heading, a threshold is used to allow the algorithm to consider the candidate whose headings are close to the sensor-detected heading. We also need to check if the segment is connected with the old segment that the vehicle was most likely traveling on in the previous computation cycle. Segments far away from the old segment should have lower priority on the list. If the vehicle was following a predetermined route during route guidance (Chapter 6), the segments along this route should be strong candidates for matching considerations. All this checking will likely identify the best matching segment. If an off-road situation (such as a parking lot) is detected, more computations are required in order to include candidate segments within a gradually growing region of the vehicle vicinity. It uses similar matching criteria to determine when and where the vehicle returns to the road network.

Finally, this best matching segment is presented to the system along with a most probable matching point on the segment. If there are no good candidates available, if the sensor-determined position coincides with the current vehicle position, or if there is an ambiguity, nothing will be reported, provided the algorithm is designed to do so. The vehicle must then rely on the sensor data alone to determine its position.

To save on computation time, a rectangular trial-and-error region rather than an error ellipse may be defined. The drawback of this method is that the error region will never be associated with the positioning errors. A statistically defined error region can be expanded and shrunk at each cycle of the algorithm while the trial-and-error region would stay at the same size all the time. As an example, we will see below how a probabilistic algorithm, which follows the basic procedures in [6] closely, works.

This probabilistic algorithm interrupts the main system program approximately once per second to provide the current vehicle position for display. During each cycle, it first updates the old dead-reckoned or map-matched position to the current dead-reckoned position, on the basis of sensor-measured data, as discussed in Section 3.2. As in the second step of Figure 4.2, the update accuracy is estimated in terms of the CEP, which defines an error region. This region encloses an area with a certain probability of containing the actual position of the vehicle.

The basic idea is to use a 3D probability density function (PDF) to define a set of equal-probability contours (Figure 4.4). Each contour is generated by an $XY$ horizontal plane slicing through the PDF at some level and encloses an area with a percentage of the probability density, such as 50%, 60%, and so on. The peak of the PDF is directly above the current dead-reckoned position. By projecting these contours onto the $XY$ plane, we obtain a set of ellipses, each with an equal probability of including the actual position of the vehicle. For instance, the 90% CEP encloses an area that has a 0.9 probability of including the actual position of the vehicle. In actual implementation, one of the ellipses is selected as the error region and is approximated by a rectangle. Reference [6] contains a set of equations used to define the four corners of the rectangle based on the error models of the fluxgate magnetic compass and the differential odometer. The area of the CEP is normally proportionately larger than the area determined by the matching point (discussed shortly) from the last cycle to reflect the accumulated error and the resulting reduction in the accuracy of the current dead-reckoned position.

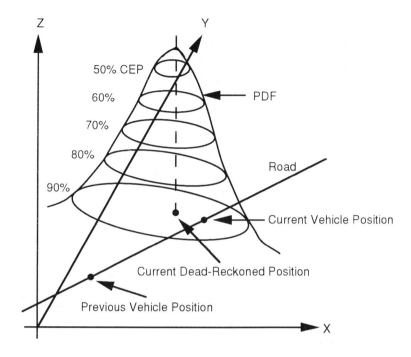

**Figure 4.4** Position error regions defined by a probability density function (PDF).

Once the error region has been defined, the algorithm performs the remainder of the steps shown in Figure 4.2, except for the last update step, to find the best matching road segment. This segment is the most likely candidate in the vicinity of the vehicle to contain a point more likely than the most recent dead-reckoned position of the vehicle to be the current vehicle position. A more detailed description for each of these steps (between the second and the last steps in Figure 4.2) has been provided in Figure 4.5.

The algorithm extracts candidate road segments (segments in the error region) by examining all segments in the neighborhood of the vehicle in sequence. The segments are fetched from the digital map database. Each segment is first tested to determine if its direction is parallel to the vehicle heading within a specified threshold. If it passes the direction test, this segment is then examined to determine if it intercepts the CEP (error region), which probably contains the vehicle. After passing these initial tests, the segment is added to the segment list. Otherwise, the algorithm returns to the top and fetches another segment for the same test provided the current segment is not the last one in the neighborhood.

The algorithm tests each segment added to the list for its connectivity and closeness. The connectivity test checks if the segment connects to the segment

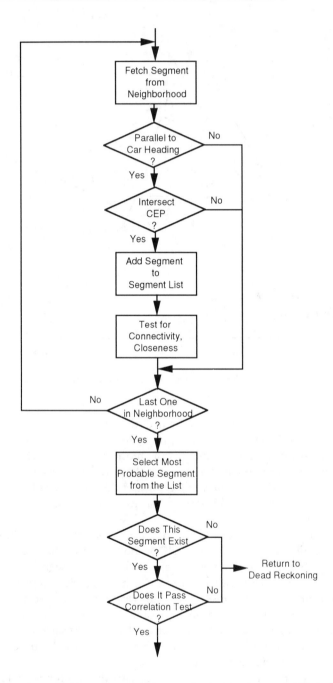

**Figure 4.5** Finding a best matching segment within a confined region in a probabilistic algorithm.

previously (last cycle) determined to contain the vehicle position. In other words, the algorithm checks if each segment on the list is connected with a previously matched segment. The closeness test checks each segment on the list to see if it is very close to any previously matched segment on which the vehicle was actually traveling. In other words, the algorithm checks to see if there is an ambiguity between any pair of segments on the list. Any segment that fails these two tests is removed from the segment list. No more than two segments on the list are permitted to survive the connectivity and closeness tests.

From the surviving segment list, the algorithm selects the most probable segment. If there is only one survivor, the task is simple. This segment is the most probable segment. If there are two segments on the list, more tests must be performed. These include a test to see which segment connects to the previously matched segment properly and which one is closer to the current dead-reckoned position. The most probable segment is selected as the winner. If these two segments have almost identical characteristics or are too close together, neither segment is selected as a matching segments, and the current dead-reckoned position is not updated.

The single most probable candidate must pass the final correlation tests to be eligible as a matching segment. If the vehicle is not turning, the algorithm performs a simple trajectory match. Otherwise, a correlation function is calculated. For trajectory matching, a short history of piecewise traveled distances and headings calculated from sensor outputs are kept in memory temporarily. These vehicle trajectory data are used to compare against the vehicle route data of piecewise segment lengths and headings from the same period. If a match exists, the sole survivor is the segment to which the current dead-reckoned position can be updated and is saved as the matching segment. For correlation function computation, the algorithm calculates the function between the vehicle trajectory and certain segments (including the most probable segment and the segments connected to it). If the results pass certain thresholds, the most probable (correlation) point on the segment derived during the calculation is saved for later use in updating the current dead-reckoned position. Obviously, the segment that contains this matching point becomes the matching segment. If this most probable candidate fails the correlation tests, the current dead-reckoned position is not updated, indicated as a rightward pointing arrow in Figure 4.5.

Last, the algorithm updates the vehicle position, which is similar to the last step in Figure 4.2. If the vehicle is turning, the vehicle position is updated to the matching point determined during the correlation function calculation. Otherwise, it is updated to a most probable constant-course position along the matching segment. Once the current position has been updated to the map-matched position, the area of the CEP will be proportionally reduced to reflect the resulting increase in the accuracy of the vehicle position. Again, see [6] for the equations used to derive the new corners of this rectangular error region. Finally, the sensors are recalibrated. If the vehicle has just finished a turn, the distance sensor is recalibrated. If the vehicle is not turning, the heading sensor is recalibrated. The turning condition of the vehicle can be easily determined either using the heading sensors or during the map-matching process. The update steps in this algorithm are shown in Figure 4.6.

We know that the probabilistic method makes no assumption that the vehicle is always on the road network. It handles off-road conditions much better than the semi-deterministic method. However, unlike traveling on a known road network, map-matching adjustments cannot prevent the accumulation of dead-reckoning errors while the vehicle is off-road. There may be considerable uncertainty in vehicle position, which could lead to incorrect conclusions when matching against the surrounding road segments. The further the vehicle travels off-road, the greater the uncertainty in the estimated position. Many possible approaches have been proposed to eliminate this problem. For instance, an absolute sensor can be used as a sanity-checking sensor to bound the absolute accuracy of the system as briefly mentioned in Section 4.4.

## 4.3 FUZZY-LOGIC-BASED MAP MATCHING

As discussed in preceding sections, the vehicle trajectory generated by a positioning module alone is often distorted or different from the actual route due to sensor and processing errors as well as imperfections of the database resulting from digitization.

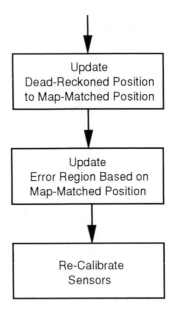

Figure 4.6 Final detailed update steps of the probabilistic algorithm.

This often causes some difficulty in the map-matching process. If a map-matching-based vehicle is traveling in a city downtown area, there may be many road patterns matching the trajectory reported by the positioning sensors at any given moment. It may be difficult to distinguish precisely on which particular street the vehicle is traveling. Rather, the system may conclude that the vehicle is "more likely" to be on certain streets, and "less likely" to be on some certain other streets. The location of the vehicle can be determined only when sufficient additional trajectory information is provided to compare against the digital map database. This ambiguity needs to be resolved if an accurate location and navigation system is desired. Therefore, techniques for dealing with qualitative terms such as likeliness are required in the map-matching process.

Fuzzy logic has been demonstrated as an effective way to deal with tasks that involve qualitative terms, linguistic vagueness, and human intervention. By making use of fuzzy sets (Appendix F), linguistic terms with vague concepts can be defined mathematically. Expert knowledge and experience is represented by a set of rules. Furthermore, inferences can be drawn through an approximate reasoning process, which simulates human reasoning capabilities. Because map matching is a qualitative decision-making process involving a degree of ambiguity, this suggests that the development of fuzzy-logic-based map-matching algorithms may be useful. Fuzzy logic can also be used in other modules or subsystems in addition to the map-matching module. Readers can refer to [9] for these potential application areas.

### 4.3.1 Fuzzy-Logic-Based Algorithms

In this section, we discuss how to apply fuzzy logic in a navigation system by constructing a simple rule set for a map-matching module. Background for readers who are not familiar with the concept of fuzzy logic can be found in Appendix F. As discussed in that appendix, a fuzzy logic inference process consists of three main parts: the fuzzifier, the inference engine, and the defuzzifier, as shown in Figure 4.7.

Our discussion of this fuzzy-logic-based algorithm follows [10,11] closely. This method is based on the approximate reasoning in fuzzy logic. It uses fuzzy logic to assign truth values from zero to one for each of the sensor signals and road segments and then make decisions based on the assigned truth values.

Figure 4.8 shows a flowchart for this fuzzy-logic-based map-matching algorithm. To save space, "road" has been used for "road segment" in this flow chart. The rules are numbered to correspond with the logic boxes in the flow chart. Some of them consist of more than one sub-rules.

As Figure 4.8 indicates, this algorithm uses eight rules. Some of the fuzzy membership functions associated with each rule are given next. In the following discussion of the rules, we assume that the root mean square error of the past five headings is used as the heading error to determine if the heading has become steady.

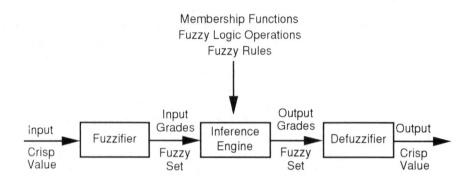

**Figure 4.7** Fuzzy logic system structure.

*Rule 1:* IF Δheading is small
THEN resemblance of the route is high

where "Δheading" is defined as the difference between the direction of the road segment and the heading of the vehicle. The membership functions "small," "large," and "high" for "Δheading" and "resemblance" are defined in Figure 4.9.

*Rule 2:* IF Δheading is almost 180° AND heading error is zero
THEN possibility of U-turn is high

where, as described in Appendix F, AND is a min (minimal) operator. Membership functions of "almost 180°" for "Δheading" and "zero" for "heading error" are defined in Figure 4.10. The membership function for "possibility of U-turn" being "high" is defined the same as "resemblance" in Figure 4.9.

*Rule 3.1:* IF Δdistance is large
THEN necessity to retrieve successive segments is high
*Rule 3.2:* IF Δheading is large and heading error is zero
THEN necessity to retrieve successive segments is high

where "Δdistance" is defined as the difference between the segment length and the distance the vehicle has traveled on that particular segment. The membership function of "large" for "Δdistance" is defined in Figure 4.11. The membership functions of "large" for "Δheading" and "high" for "necessity" (same as "resemblance") are defined in Figure 4.9.

*Rule 4:* IF the heading errors and the root mean square errors for the vehicle speed are small
THEN the motion is steady
*Rule 5:* IF the truth value of the previous candidate road pattern is high
AND

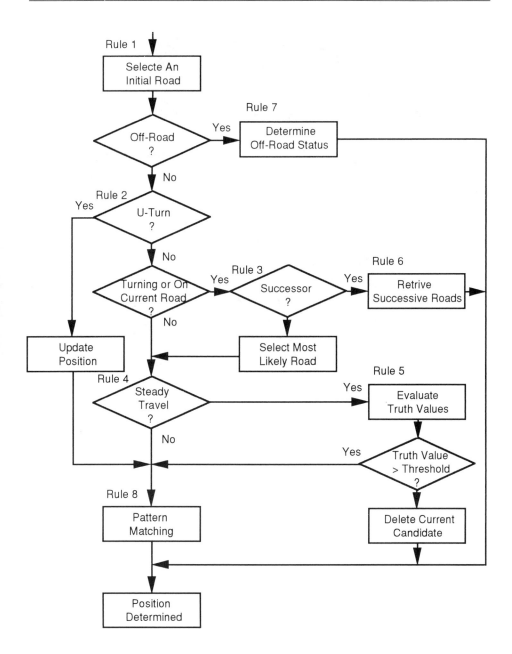

**Figure 4.8** A fuzzy-logic-based map-matching algorithm.

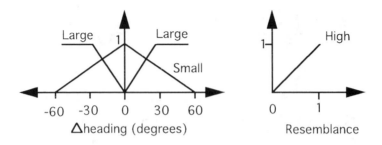

Figure 4.9  Membership functions for Rule 1.

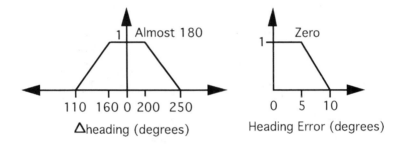

Figure 4.10  Membership functions for Rule 2.

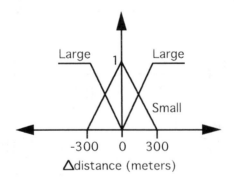

Figure 4.11  Membership function for Rules 3.

IF the truth value of the current candidate road pattern is high
THEN the truth value of the correspondence is high

*Rule 6.1:* IF the difference between the distance traveled along the current candidate road and the length of the candidate road is small
AND

IF the difference between the vehicle heading and direction of the successive road is small

THEN the truth value for the successive road is high

*Rule 6.2:* IF the truth value of the candidate road is high

AND

IF the truth value for the successive road is high

THEN the combined truth value of the moving vehicle on the successive road is high

*Rule 6.3:* IF no road pattern similar to the path of travel can be found within a given distance

THEN the vehicle is off-road

*Rule 7:* IF there is more than one road pattern within a given distance similar to the current vehicle motion

THEN the vehicle is on-road

*Rule 8.1:* IF $\Delta$heading is small AND $\Delta$distance is small

THEN resemblance of this segment is high

*Rule 8.2:* IF resemblance of this segment is high AND resemblance of history path is high

THEN resemblance of the whole path is high

where the AND operator in Rule 8 reflects the different degree of resemblance between successive segments and the vehicle trajectory. The membership functions of "small" for "$\Delta$heading" and "$\Delta$distance" are defined in Figures 4.9 and 4.11, respectively.

Through this matching process, a candidate road segment may be removed due to its resemblance being too low, for example, zero. On the other hand, a new candidate segment may be generated if the vehicle is either on or close to another segment. Because the resemblance is constantly being evaluated, the matching possibility for each candidate segment will decrease since continuous evaluation causes the truth value of the old candidate segment to decrease as the vehicle keeps moving. One way to continue the process is to reset the degree of resemblance to one whenever there is only one candidate.

The preceding set of rules is primitive. Implementation of a robust fuzzy-logic-based map-matching module requires the use of additional conditions, such as drivability of successive road segments, or the addition of special rules for GPS, if a GPS receiver is used. However, this example demonstrates very well how fuzzy logic can be successfully used in a map-matching algorithm. With these primitive rules as a foundation, one can easily develop a good map-matching module. Many fuzzy-logic development tools are available. New rule sets can be developed and tested using these tools. Users can either use the tool to generate the embedded computer language code directly for integration with the system, or design their own fuzzy inference engine to be embedded in the system [12].

The basic principles of the conventional and fuzzy-logic-based map-matching algorithms are the same. Both algorithms try to identify the road segment on which the vehicle is traveling and precisely determine the position of the vehicle on this segment to update the dead-reckoned position. The main difference is that one approach determines this matching segment by probabilistic reasoning and the other does it via a fuzzy-logic inference process. More specifically, the conventional algorithm defines the sensor error models relative to the candidate segments in probabilistic terms. The fuzzy-logic-based algorithm, on the other hand, uses membership functions to describe the sensor error models in relation to the candidate segments. The matching processes for both algorithms are essentially the same. A possibility is associated with each candidate segment. These possibilities increase or decrease according to similarity. Candidates with low possibilities are discarded and the one with the highest possibility is presented to the system. However, different methods can be used to evaluate the possibility. This leads to the interesting conclusion that other inference methods besides the two methods discussed here can also be used in map-matching algorithms.

## 4.4 OTHER MAP-MATCHING ALGORITHMS

Map matching is basically a pattern recognition process. Therefore, many pattern recognition methods could be used. One such candidate method is based on a neural network. As mentioned in Section 3.5, neural networks consist of many layered nodes (neurons) that are interconnected. A neural network is a dynamic system and must be trained to rearrange the layers and interconnections to model the real-world application. Besides different pattern recognition methods, centralized, decentralized, or hybrid approaches can be used to work with positioning sensors. In the centralized approach, the integrated and fused multisensor data are matched with the map. In the decentralized approach, each individual sensor or group of sensors is matched with the map separately. In the hybrid approach, the centralized and distributed methods are applied sequentially to obtain a better matching result. Although there are many variations on this technique, the underlying principle is the same: A digital map is used to filter out vehicle sensor errors and to determine the best position. As examples of these variations, we show some typical map-matching methods used with GPS receivers in Figure 4.12.

Because of the popularity of GPS receivers, many applications have been developed on the basis of this satellite-based radio navigation technology. There is no question that a map-matching process can also help improve the accuracy of the receiver output as shown in Figure 4.12(c), especially, when selective availability (SA) is on (see Section 3.4.2).

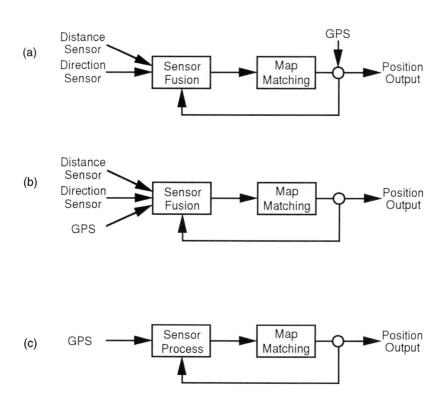

**Figure 4.12** Variations of map-matching algorithms used with GPS: (a) as a sanity-checking sensor, (b) as one of integrated sensors, and (c) as the only sensor.

## 4.5 MAP-AIDED SENSOR CALIBRATION

Map data elements (attributes) can be used to calibrate both distance sensors and angular sensors. These may be done during a trip as briefly discussed for certain sensors in Chapter 3. For instance, sensor blunders (short-term anomalies) can be detected and corrected using the map-matching results [13]. The length of a road segment can be compared against the distance measured by the distance sensor. The direction of a road segment can be compared against the direction measured by the direction sensor. When the vehicle travels on this segment, any errors above a certain threshold will be detected and the corresponding corrections and calibration can be performed.

Assume that the map database is accurate and the fact that the vehicle is traveling on a specific road segment can be confirmed by a map-matching-based positioning module. The system can record the starting and ending points of the segment. If the segment is long enough, a comparison can be done to monitor the performance of the distance sensor and the direction sensor. For instance, the length of the segment should be nearly equal to the distance traveled by the vehicle on the segment and the direction of the segment should be closely equal to the heading of the vehicle as it travels along that segment. Otherwise, either a sensor anomaly (blunder) has occurred or the scale factor of the sensor needs to be recalibrated. Recall that many sensors could be affected by the environment so that recalibration becomes necessary. Using a map, this can be done dynamically during the journey.

A digital map database can help a dead-reckoning-based positioning module present very good position information to the location and navigation system. When developing a map-matching-based positioning system, we must keep in mind that this system relies heavily on the map. Therefore, the accuracy of the vehicle position determined by the system should be similar to that of the map database itself. If the map is of low quality, the system may not meet expected design specifications or user requirements.

As mentioned, a dead-reckoning system has a tendency to accumulate sensor errors, so that the system may eventually fail to locate its position. A map database may be unable to provide accurate information due to unmapped roads, off-road travel, human errors during the mapping and conversion process, etc. Integration of dead-reckoning and map-matching systems may improve the performance of the positioning module. However, it still may fail due to the fact that the errors are accumulative and/or the presence of error in the maps. User intervention may be required to recover. Since the GPS satellite deployment is now complete, it is very common to find location and navigation systems that integrate GPS with dead reckoning and map matching to boost the self-recovery ability, and to provide system reliability. Besides, as a source of independent position, direction, and velocity information for the vehicle, GPS can help to calibrate the dead-reckoning sensors and detect sensor blunders dynamically. On the other hand, GPS receivers can occasionally generate suspicious position and velocity information or even no solution at all because of satellite signal blockage and multipath reflection. Dead-reckoning sensors and the map can detect GPS blunders and fill voids by erratic satellite signals. As we shall see in Section 10.2, similar map-matching techniques can also be used to assist terrestrial-radio-based positioning or other systems. By complementing each other, these systems provide much better performance than systems that do not use these matching techniques (Table 10.1).

In conclusion, let us consider an interesting question. During the segment matching process, how can you determine which is the best matching candidate when a road forks into almost parallel roads with very small differences in road direction?

### References

[1] R. L. French, "Map Matching Origins, Approaches and Applications," *Proc. Second International Symposium on Land Vehicle Navigation*, July 1989, pp. 91–116.

[2] A. Papoulis, *Probability, Random Variables, and Stochastic Processes*, 3rd ed., New York: McGraw-Hill, 1991.

[3] B. Porat, *Digital Processing of Random Signals: Theory and Methods*, Englewood Cliffs, NJ: Prentice Hall, 1994.

[4] H. L. Van Trees, *Detection, Estimation, and Modulation Theory*, New York: Wiley, 1968.

[5] Motorola, *Oncore User's Guide*, Northbrook, IL: Position and Navigation Systems Business, May 1996.

[6] S. K. Honey, W. B. Zavoli, K. A. Milnes, A. C. Phillips, M. S. White, and G. E. Loughmiller, Jr., "Vehicle Navigational System and Method," *United States Patent No. 4796191*, Jan. 1989.

[7] M. L. G. Thoone, "CARIN, A Car Information and Navigation System," *Philips Tech. Rev.*, Vol. 43, No. 11/12, Dec. 1987, pp. 317–329.

[8] M. Cyrus and J. Beck, "Generalized Two- and Three-Dimensional Clipping," *Computers & Graphics*, Vol. 3, 1978, pp. 23–28.

[9] Y. Zhao, A. M. Kirson, and L. G. Seymour, "Fuzzy Logic Applications in Vehicle Control and Navigation," *Abs. Third World Congress on Intelligent Transport Systems*, Paper #1225 (full paper in CD-ROM), Oct. 1996, p. 204.

[10] L.-J. Huang, W.-W. Kao, H. Oshizawa, and M. Tomizuka, "A Fuzzy Logic Based Map-Matching Algorithm for Automotive Navigation Systems," *IEEE Roundtable Discussion on Fuzzy and Neural Systems, and Vehicle Applications*, Paper No. 16, Nov. 1991.

[11] W.-W. Kao and L.-J. Huang, "System and Method for Locating a Traveling Vehicle," *United States Patent No. 5283575*, Feb. 1994.

[12] G. Viot, "Structuring Fuzzy in C," *Proc. 7th Annual Embedded Systems Conference*, Sep. 1995, pp. 153–170.

[13] Y. Zhao, L. G. Seymour, and E. M. Kozikaro, "Absolute Heading Sensor Blunder Detection Using Relative Heading Sensor and Road Segment," *Technical Developments*, Motorola, Vol. 29, Nov. 1996, pp. 104–107.

# CHAPTER 5

▼▼▼

# ROUTE-PLANNING MODULE

## 5.1 INTRODUCTION

Route planning is a process that helps vehicle drivers plan a route prior to or during a journey. It is widely recognized as a fundamental issue in the field of vehicle navigation. Route planning can be further classified into either multivehicle (system-wide) route planning, which plans multidestination routes for all vehicles on a particular road network, or single-vehicle route planning, which plans a single route for a single vehicle according to the current location and a given destination.

In the computer literature, people often refer to finding a route from point A to point B as a *shortest path* problem. Many algorithms have been developed to solve single-origin shortest path and all-pairs shortest path problems [1]; these algorithms treat situations that are quite analogous to single-vehicle route planning and multivehicle route planning, respectively. In this chapter we discuss mainly algorithms for solving the single-vehicle route planning problem. Multivehicle route planning is more complex than single-vehicle route planning. However, the background on solving the single-vehicle route-planning problem will facilitate the study of the multivehicle route-planning case. Readers interested in multivehicle route planning will find a brief discussion in Section 5.6 as well as in relevant readings on dynamic programming, operations research, data structures, and algorithms [1–3].

In this chapter, we follow the convention used in Chapter 2. We define a *node* to be a road intersection or a dead-end point on a road, and a *segment* to be a piece

---

of roadway between two nodes. Note that segments are often called *edges, arcs,* or *links,* and nodes are often called *vertices* or *points* in the computer literature on directed graphs.

A variety of route optimization criteria (planning criteria) may be used in route planning. The quality of a route depends on many factors such as distance, travel time, travel speed, number of turns and traffic lights, and dynamic traffic information. We refer to all these factors as the *travel cost.* Some drivers may prefer the shortest distance. Others may prefer the shortest travel time, etc. Therefore, the evaluation function chosen to minimize this cost depends on the system design and user preference. These route selection criteria can be either fixed by a design or implemented via a selectable user interface. To minimize the travel distance (road segment length), distance can be stored in a digital map database, so that the route-planning algorithm can use the database when performing minimization. If travel time is being minimized, the road segment length and speed limit can be stored for use in calculating the travel time for each road segment. As noted in Section 2.6, segment length and speed limit are often defined as attributes of the road segment in the digital map database. In summary, determination of the best route involves using a digital map to select a route that minimizes variables such as time and distance.

Before introducing route-planning algorithms, we shall familiarize ourselves with some commonly used technical terms from computer science. To analyze the performance of various algorithms, we need to know the key factors that affect performance, and the common terminology used to discuss them. The *running time* (execution time) of a program depends on the input to the program, the quality of code generated by the compiler used, the nature and speed of the machine instructions, and the time complexity of the algorithm. The running time of a program is defined as a function of the input. For many programs, the running time is a function of some particular input. Often, this computation time depends not on the exact input but only on the "size" of the input. We usually define this time complexity to be the worst-case running time. Because the running time of a program depends on the compiler used to compile the program and the machine used to execute it, we cannot express the time in standard units such as seconds. Instead, all we can say is that the running time of a certain algorithm is proportional to the size of the input, etc. The constant of proportionality is left unspecified since it depends very heavily on the compiler, the machine, and other factors.

*Big O notation* is commonly used to describe algorithm running times. For instance, let the size of the input to the algorithm be $n$. We thus use $O(1)$ to mean computing time is approximately independent of the size of the input, or bounded by a constant. The term $O(n)$ means that the running time is directly proportional to $n$, and is called *linear time.* The term $O(n^2)$ refers to *quadratic time;* $O(n^3)$ is called *cubic time;* $O(c^n)$ is called exponential time, where $c$ is a constant; and $O(\log n)$ and $O(n \log n)$ are called logarithmic time. Thus, programs can be evaluated by comparing their running time functions (neglecting constants of proportionality). In the preceding notation, for instance, a program with running time $O(n)$ is better than one with running time $O(n^2)$ even though they are both polynomial-time algorithms. If you are not interested in the timing analysis, you should be able to skip the sentences and paragraphs discussing this subject and still understand the algorithms.

The timing analysis is often referred to as the *time complexity*. Similarly, the analysis of memory used to perform the search is called the *space complexity*. The *completeness* of an algorithm refers to the ability of an algorithm to guarantee obtaining a solution if one exists. If the algorithm is capable of identifying the best of several different solutions, it is referred to as being *optimal*. In this chapter, we focus on the time complexity to give readers a basic idea of how to compare various planning (or search) algorithms. We now discuss several popular algorithms. The interested reader may refer to the subject under search in [4] to see how some of the algorithms discussed here can be implemented in artificial intelligence and other computer languages.

## 5.2 SHORTEST PATH

We first study a popular shortest path algorithm, and then learn how to modify it to solve the route-planning problem.

### 5.2.1 Dijkstra's Shortest Path Algorithm

One of the main representatives of the shortest path algorithm is Dijkstra's algorithm [5,6]. This algorithm uses a greedy technique often employed in optimization problems. Greedy algorithms make locally optimal choices at each step in the hope that these choices will produce a globally optimal solution. Given a weighted and directed graph, similar to Figure 5.1, Dijkstra's algorithm maintains a set of nodes (e.g., small circles with node numbers in them) whose minimum costs (e.g., labeled weights between nodes) relative to the origin node are already known. Initially, this set contains only the origin node, with all other nodes being "remaining nodes." At each step, we add to the set one remaining node whose cost relative to the origin node is as small as possible. Assuming all segments have non-negative costs, we can always find a shortest path route from the origin node to the remaining node that passes only through nodes in the set. Call such a route "special." At each step in the algorithm, we use an array to record the length of the shortest route to each node. Once the set includes all nodes, all routes are "special," so this array will hold the shortest distance from the origin to each node. This algorithm bears some similarity to the breadth-first search algorithm, which is roughly analogous to

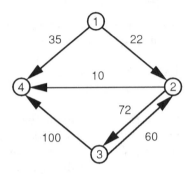

**Figure 5.1** Directed graph with costs.

level-by-level traversal of an ordered tree [1,7]. It is easy to convert a road network to a directed graph similar to Figure 5.1.

One necessary condition for using this algorithm is that the cost must be non-negative for each directed segment, which is always true in vehicle navigation. A directed segment means that each road segment has a travel direction associated with it. Provided this non-negative cost condition is satisfied, Dijkstra's algorithm will find optimal solutions from the origin node to every other node. In our example, this means that this algorithm will find minimum-cost routes from node 1 to nodes 2, 3, and 4, respectively.

Let $n$ stand for the total number of nodes and $s$ for the total number of segments in the road network. The running time for the entire Dijkstra algorithm is $O(n^2 + s) = O(n^2)$. The equality results from the fact that we can always find constants $c_1$, $c_2$, and $c_3$, which satisfy $c_1 n^2 + c_2 s \le c_3 n^2$, and constants are always ignored in Big O notation. Therefore, the running time of Dijkstra's algorithm is quadratic.

Let $b$ stand for the *branching factor*, which is the average number of segments starting at each node. If the road network is sparse, that is, $b$ is much less than $n$ ($b \ll n$), it is practical to organize the nodes into a priority queue with a binary heap [1]. The resulting algorithm is called a modified Dijkstra algorithm. For vehicle navigation problems, it is easy to satisfy $b \ll n$ because the number of intersections in the problem domain is usually much larger than the average number of streets entering each intersection. The total running time of this algorithm is $O((n + s)\log n)$ or $O(s \log n)$ if all nodes are reachable from the origin node. Another modification is to implement the priority queue with a Fibonacci heap. As a result, the total running time spent in this version of Dijkstra's algorithm is $O(n \log n + s) = O(n \log n)$. In other words, using better data structures reduces the quadratic running time of the Dijkstra algorithm to logarithmic time. Further reduction of the running time has also been reported by using other data structures such as a radix heap [8].

### 5.2.2 Modified Shortest Path Algorithm

For single-vehicle navigation, it is not necessary to find the optimal route from the origin node to every other node in the road network. Dijkstra's algorithm can be modified to terminate when it has found the optimal route to the destination node. Listed later is a double linked-list implementation of the modified shortest path algorithm (or uniform-cost search [9]). Each node in the list has a back pointer that points to its predecessor (parent) node and one or more next pointers that point to its successor (children) nodes or a null. Although this algorithm is not very popular for route planning, it is still beneficial to discuss it, so that the reader can compare it with other algorithms and become familiar with the associated terminology. We will see that the other algorithms discussed in this chapter are basically further improvements of this algorithm. Note that in general there are a variety of implementations for each algorithm. When studying these algorithms, try to understand the algorithm itself and how it can be used to solve the route-planning problem rather than any particular implementation technique.

As mentioned in Chapter 2, the road network is represented by a digital map database. A route-planning algorithm needs to operate by searching through this directed road network, in which each node represents an intersection or a dead-end point of a road. Assume that in the algorithm we convert the node in the map database into a node data structure containing at least $x$ and $y$ coordinates, current planning status, and some road attributes, an indication of its promise as an intermediate routing point, a parent link that points back to the "best node" from which it came, and a next link that points to the next node in one of two linked lists, OPEN and CLOSED:

OPEN: a list of the nodes that have been generated but not yet expanded.
CLOSED: a list of the nodes that have been expanded.

The term *generate* means to create a data structure corresponding to a particular node, and the term *expand* means to generate all successors (children) of a node. This data structure should include all information required for planning as discussed briefly in the last paragraph. Figure 5.2 shows an example of generated and expanded nodes on the respective lists, where $b$ and $c$ are successors of $a$. A black dot indicates the head of each list. An electrical ground signal indicates the end of the list. In the actual implementation, a pointer to null is used. To simplify the figure, the nodes have been represented using small circles, and the data structure and pointer arrows associated with each node have been ignored. The letters in each circle indicate the name of the list, OPEN or CLOSED.

An evaluation function (cost function) is used to estimate the merit of each node generated. This function enables the algorithm to search the more promising routes first. The evaluation function is defined for node $n$ as $g(n)$, where $g(n)$ is a measure of the actual cost of getting from the origin node to the current node $n$. It

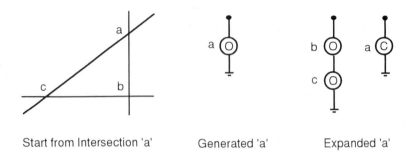

Start from Intersection 'a'          Generated 'a'          Expanded 'a'

**Figure 5.2** Linked lists with generated and expanded nodes for road intersections.

is up to the system designer or user to determine the travel cost used for the evaluation function as discussed in Section 5.1.

### Modified Shortest Path Algorithm

1. Let the initial OPEN list contain only the origin node with zero cost ($g$ value) and the CLOSED list be empty. Let the costs of all other nodes be infinity.
2. If there are no nodes on the OPEN list, report failure. Otherwise, select the node on the OPEN list with the lowest cost ($g$ value) and call it BEST. Remove it from the OPEN list and place it on the CLOSED list. See if BEST is a destination node. If so, go to step 3. Otherwise, generate the successors of BEST based on its connecting segment attributes contained in the map database. For each such successor node $n$, perform the following steps:

   2a. Compute the cost of $n$: $g(n)$ = cost of BEST + cost from BEST to $n$.

   2b. If $n$ matches a node already in OPEN, check to see if node $n$ has a lower cost ($g$ value). If so, replace the matching node's cost with the cost of node $n$, and set the matching node's back pointer to BEST.

   2c. If $n$ matches a node already in CLOSED, check to see if node $n$ has a lower cost ($g$ value). If so, replace the matching node's cost with the cost of node $n$, set the matching node's back pointer to BEST, and move the matching node to OPEN.

   2d. If $n$ is not already in either OPEN or CLOSED, set node $n$'s back pointer to BEST and place node $n$ in OPEN. Repeat step 2.

3. From BEST, traverse the back pointer up to the origin node and report the solution route.

This algorithm will find a "best" route from a given origin to a given destination. The same techniques used in Dijkstra's algorithm can be used to reduce the running time of the modified shortest path algorithm. Another interesting fact is that if all segments have equal cost, the algorithm becomes a breadth-first search. In our discussion, we have ignored the case where an origin or a destination may not coincide with any node on the map. Every node is associated with a latitude and a longitude as discussed in Chapter 2. The user interface and positioning system can help determine the latitude and longitude of the origin and destination. We can then find the closest node in the map database to either the origin or destination without much difficulty.

In this algorithm, we check each successor node to see whether it is on either the OPEN or CLOSED list already in order to avoid having a node appear more than once. It is possible that more than one route may reach a destination passing through a same particular node or intersection. The fact of a node encountered more than once during the search, i.e., a recent successor node is already on the OPEN or the CLOSED lists, reflects in reality that multiple routes exist via this node. One of these routes should have a minimal cost. This check guarantees that the optimal route is always kept in memory.

Another technique often used in route planning is to include forbidden states (or drivability flags) in the data structure for each node [10]. In general, any road network will contain a fair number of dead-end roads, blocked roads, or one-way roads. We cannot use them as traversable roads for route planning. The same is true for traversing a one-way road in the wrong direction. A simple way of dealing with this problem is to dynamically incorporate this information into forbidden states for the corresponding node data structures. Forbidden nodes cannot be used in construction of the search tree, so the system automatically avoids forbidden roads. One alternative is to avoid generating a successor to a valid node for the OPEN list if the node is connected to a forbidden segment. We can further expand this idea to enhance the route-planning function. For instance, a user may want to avoid expressways for a particular trip. If the system is designed in such a way that the user can make this choice before executing the planning algorithm (one of the planning criteria), the system can then mark all expressways as forbidden to satisfy the user's request.

There is one difference between the usage of the shortest path algorithm and usage of the modified version described above in vehicle navigation applications. The shortest path algorithm is often used to find the solution for a given graph, or to precompute routes off-line for storage in memory (see Section 5.6 for examples), while the modified version and other algorithms are used when real-time navigation is required. When a navigation system is up and running, the user expects the best route to be presented almost instantaneously. Although we can view a large, detailed road network as a graph from the vehicle-navigation point of view, it is impractical (given current technology) to load the entire road network into main memory for route planning. The common practice is to load, on demand, the partial network likely to be used during the planning process into memory. We do not have the entire graph to work with, especially during real-time computation. We can imagine, then, that we are searching a tree with the origin node as the root and the rest of the generated nodes as leaves or descendants (see the example tree shown later in Figure 5.5). Therefore, a different measurement function is needed to evaluate the running time of the algorithm.

Let $d$ stand for the *search depth* of the shortest solution route from the origin node to a destination node. The search depth is the number of levels in a tree a search must traverse in order to find a solution. For instance, if we treat the root as level 0, all successors (leaves, children) of the root are at level 1 of this tree. In this notation, the running time of the modified shortest path algorithm becomes $O(b^d)$. On completion of the planning process, the total number of nodes on the tree in main memory should be approximately equal to $b^d$.

The modified shortest path algorithm takes more time and space to execute than a heuristic search algorithm. In the next section, we introduce a heuristic search algorithm. A simple example is discussed, and the final search tree, final OPEN list, and final CLOSED list are shown so that potential users can become more familiar with this popular algorithm.

## 5.3 HEURISTIC SEARCH

A heuristic search is an informed search strategy, which means that the search has information concerning the number of steps or the cost from the initial state and current state to the destination state. Examples of informed search methods include the best-first search, the memory bounded search, and iterative improvement algorithms such as the hill-climbing search and simulated annealing [11].

### 5.3.1  A* Algorithm

The most popular heuristic search used in route planning is the A* algorithm [12], which is a best-first search method. The A* algorithm has been used in fields other than vehicle route planning, such as robotics [13] and other areas where minimum-cost solutions are required.

The reason that we must do a search is that we do not know exactly which way to go. The situation would be much better if we knew in advance. Therefore, heuristic information is introduced to improve the efficiency of the search process. A heuristic is a rule of thumb, a technique for improving the efficiency of a search process by possibly sacrificing claims of completeness [14]. Heuristic information can help to determine which is the "most promising" node, which successors to generate, and which irrelevant search branches to prune. The purpose of using a heuristic is to provide an estimate of how far a node is from the destination node so that the system can determine how likely that particular node is to be on the best solution route. Heuristics are good to the extent that they point in the right direction, but they may temporarily or occasionally lead us to search an inappropriate node, such as a dead-end street, for a particular trip. Despite this temporary problem, the process will still lead to the destination node if such a route exists. Using good heuristics may enable us to obtain good solution routes in less time using less memory space.

The A* algorithm is an extremely useful technique for implementing a best-first search of a road network. If the evaluation function selected satisfies a certain property, explained below, the A* algorithm is guaranteed to find an optimal route, if one exists, while searching less space than Dijkstra's algorithm or the modified shortest path algorithm [15,16]. To understand this savings, we use a circle to represent roughly the abstract space expanded in the modified shortest path algorithm (or Dijkstra's algorithm), and an egg-shaped area to represent roughly the abstract space expanded in the A* algorithm. Thus, the space examined in the two algorithms can be represented as shown in Figure 5.3.

A good way of using heuristic information is to calculate a heuristic function that evaluates every node generated to determine its goodness or badness. In this way, having the heuristic function suggest which route to traverse first when more than one route is available will make the search as efficient as possible. In the A* algorithm, this heuristic evaluation function enables the algorithm to search the most promising nodes first. The evaluation function $f'(n)$ for node $n$ is defined as:

$$f'(n) = g(n) + h'(n)$$

where $g(n)$ is a measure of the actual cost of getting from the origin node to the current node $n$, and $h'(n)$ is an estimator for the minimum cost to get from the current node $n$ to the destination node. Note that the primes in this equation represent

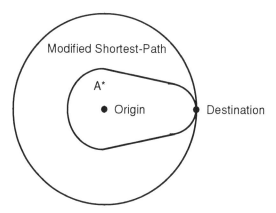

**Figure 5.3** Space examined in the modified shortest path and A* algorithms.

functions that contain estimated costs. Comparing the two evaluation equations discussed so far, we see that if $h'(n) = 0$ for $f'(n)$, this equation reduces to the one used by the modified shortest path algorithm.

As before, it is up to the system designer or user to determine the cost used in the evaluation function. For instance, if the shortest travel time is preferred, one choice is to adopt the following equation for the heuristic evaluation function:

$$f'(n) = t'(n) = g(n) + h'(n) = \sum_{i=1}^{n} \frac{d_i(n)}{v_i(n)} + \frac{d'(n)}{v'}$$

where $t'(n)$ is the travel time, and $d_i(n)$ is the actual travel distance for segment $i$ and $v_i(n)$ is the maximum travel speed (speed limit) for this segment. The actual measure $g(n)$ is the sum of the travel time from the origin node to the current node on a segment-by-segment basis. The travel time for each segment is obtained by dividing the actual distance $d_i(n)$ by the maximum travel speed $v_i(n)$ on the segment. The estimator $h'(n)$ is the Euclidean distance $d'(n)$ between the current node $n$ and a given destination divided by the estimated maximum travel speed $v'$. Note that we use the Euclidean distance as our distance estimator. This distance is the shortest distance from the current node to the destination, which is always less than or equal to the actual distance. If the estimated maximum travel speed is greater than or equal to the actual travel speed for any segment, we will always underestimate the travel time. In short, the objective of the algorithm is to minimize $f'(n)$. In this example, the travel time is being minimized, that is, $\min(t'(n))$.

The operation of the algorithm proceeds in steps, expanding one node (beginning at a given origin node) at each step, until the node that corresponds to a given destination is generated. At each step, only the most promising (the highest priority, i.e., the "best") of the generated nodes is expanded and closed, and the successors of the "best" node are opened. The "best" node is the one having the minimum cost, $\min(f'(n))$. During this process, nodes that are associated with restrictions such as a segment running in the wrong direction or with a dead-end node are excluded so that nodes leading to illegal turns or dead-end streets are not considered in the search or included in the final route. The heuristic function is applied to evaluate how promising each expanded node is after checking to see if any expanded nodes have been generated before. This check guarantees that each node appears only once on either of the two linked lists. The above procedure is repeated until the destination node is reached, at which point, the nodes currently linked from the destination node to the origin node are reversed and used as the solution route. The resultant list of nodes and associated segments or the augmented list of them is often called the maneuver list, because the vehicle will later need to follow this list in order to maneuver through the road network to reach the desired destination.

If the heuristic function $h'(n)$ never overestimates the actual cost of reaching the destination node, A* will find the optimal solution for reaching the destination.

This is called the *admissibility property*. As we see from our sample evaluation function, overestimation of $h(n)$ can easily be avoided because the function $h'(n)$ underestimates the cost for any node $n$. Therefore, provided the heuristic function is carefully selected, the admissibility property is readily achieved. The A* algorithm can generally be shown to be optimal, because it generates the fewest nodes in the process of finding the solution to a problem. Since all route costs and cost estimates are positive because of the shortest-route cost criterion and the admissibility property, the algorithm will find an optimal route for the vehicle, provided that one exists.

## A* Algorithm

1. Let the initial OPEN list contain only the origin node with zero cost ($g$ value) and the CLOSED list be empty. Let the costs of all other nodes be infinity.
2. If there are no nodes on the OPEN list, report failure. Otherwise, select the node on the OPEN list with the lowest cost ($f'$ value) and call it BEST. Remove it from the OPEN list and place it on the CLOSED list. See if BEST is a destination node. If so, go to step 3. Otherwise, generate the successors of BEST based on its connecting segment attributes contained in the map database. For each such successor node $n$, perform the following steps:
   - 2a. Compute the cost of $n$: $g(n)$ = cost of BEST + cost from BEST to $n$.
   - 2b. If $n$ matches a node already in OPEN, check to see if node $n$ has a lower cost ($g$ value). If so, replace the matching node's cost with the cost of node $n$, and set the matching node's back pointer to BEST.
   - 2c. If $n$ matches a node already in CLOSED, check to see if node $n$ has a lower cost ($g$ value). If so, replace the matching node's cost with the cost of node $n$, set the matching node's back pointer to BEST, and move the matching node to OPEN.
   - 2d. If $n$ is not already in either OPEN or CLOSED, set node $n$'s back pointer to BEST, place node $n$ in OPEN, and compute the heuristic evaluation function (cost) for node $n$: $f'(n) = g(n) + h'(n)$. Repeat step 2.
3. From BEST, traverse the back pointer up to the origin node and report the solution route.

For completeness, we have described the entire algorithm. On carefully examining the A* algorithm, we see that the only difference between this algorithm and the modified shortest path algorithm is that the evaluation function has been a heuristic one. To highlight the difference, we have marked it with a vertical bar in step 2d. The modified shortest path algorithm is actually a special case of the A* algorithm with $h'(n) = 0$.

To understand the algorithm better, let us see how the A* algorithm builds a search tree for the simple road network shown in Figure 5.4, where each intersection is represented by an English letter.

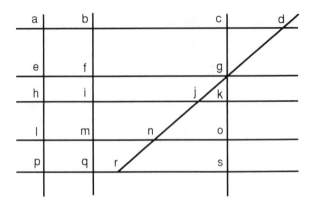

**Figure 5.4** A simple road network.

For this trip, assume that the origin is at *m* and the destination is at *g*. On completion of the route-planning process, the final search tree will appear as shown in Figure 5.5. As before, we use a small circle to represent each node, while ignoring the data structure and pointer arrows associated with the nodes. We use a straight line to represent both the upward pointers (back pointer) pointing from a node to its predecessor when applicable and the downward pointers (next pointer) pointing down to the successors for a node. The letters in the circles imply that a node on the search tree is either in an OPEN list or in a CLOSED list. Although each node on the tree could appear more than once as a successor, it can occur only once on any one of the lists, since the checking steps in 2b and 2c eliminate the redundant nodes. This multiple appearance of nodes in the tree is not difficult to understand,

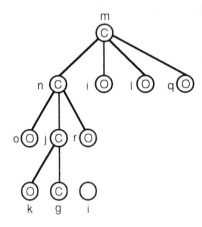

**Figure 5.5** Final search tree for the A* algorithm.

since there might be different routes to the destination, which pass through the same intersection. Note that this did not occur in our simple example. What does occur is that the algorithm never checks the final successor *i* and places it on the tree because it has already found the destination node *g*. On traversing the series of back pointers, the best route turns out to be *m-n-j-g*.

The final OPEN and CLOSED lists for this example are shown in Figure 5.6. As before, a black dot indicates the head of the list and an electrical ground signal indicates the end of the list. In the actual implementation the end of the list is indicated by a pointer to null. The OPEN list is arranged in such a way that the minimum-cost node is always at the top of the list.

As before, let *b* stand for the branching factor, which is the average number of segments starting at each node. Let *d* stand for the search depth of the shortest solution route from the origin node to the destination node; A* algorithm has running time $O(b^d)$. An initial glance suggests that the running time in Big O notation is the same as that for the modified shortest path algorithm. In reality, the heuristic information used in the A* algorithm will in general reduce the time complexity, because the value of *b* for the A* algorithm should be smaller than that for the modified shortest path algorithm. After all, this is the main reason that we are using the heuristic approach. As in the modified Dijkstra algorithm, the running time of the A* algorithm can be further reduced by using better data structures.

Because of the increasing popularity of research and development in the field of vehicle navigation, a large number of experimental studies are available for various route-planning algorithms [17–19]. There is no doubt that these activities will continue because route planning is one of the key components of a vehicle navigation system.

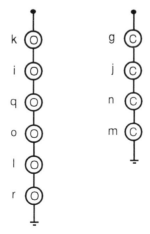

**Figure 5.6** Final OPEN and CLOSED lists for the A* algorithm.

## 5.4 BIDIRECTIONAL SEARCH

Our understanding of the shortest path and heuristic search algorithms will now enable us to generalize these two algorithms to obtain two bidirectional search algorithms. In our discussion of the previous algorithms, we always assumed that the algorithm searched for the minimum-cost route from a given origin node to a destination node. This is called a *forward search*. A *backward search* from the destination node to the origin node should actually yield the same results, as long as the cost for each segment is the same. This leads to the idea of a bidirectional search.

The idea of a bidirectional search algorithm was first proposed in [20], formalized in [21], and further analyzed and expanded in [15,22]. The basic procedure for a bidirectional search is to compute the minimum-cost route from the origin node to the destination node (forward search) at the same time as the minimum-cost route from the destination node to the origin node (backward search). If the search algorithm is running on a single processor, the algorithm must switch quickly between the forward and backward searches. A bidirectional search algorithm may reduce the search space by at least 50%, so that the running time of this algorithm may be much shorter than that of a unidirectional search. The general relationship between the search space examined by the modified shortest path algorithm and that examined by the bidirectional search algorithm is shown in Figure 5.7.

Successful implementation of the bidirectional search algorithm requires two additional conditions (assuming a single processor is used): first, a criterion for stopping the search, and second, a criterion for alternating between the forward and backward searches. For instance, we can define a stopping criterion stating that the first node must be reached from both search directions and define a static alternation

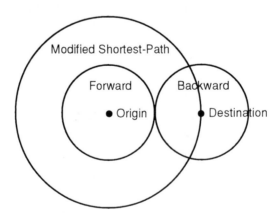

**Figure 5.7** Search spaces examined by the modified shortest path and bidirectional search algorithms.

criterion requiring 10 iterations in each direction. If these criteria are constructed properly, the bidirectional search will usually examine fewer nodes and compute a solution in less time than the unidirectional search.

*Bidirectional Search Algorithm*

1. Let the initial forward OPEN$_f$ list contain only the origin node with zero cost ($g$ value) and the forward CLOSED$_f$ list be empty. Let the costs of all other nodes be infinity.
2. Let the initial backward OPEN$_b$ list contain only the destination node with zero cost ($g$ value) and the backward CLOSED$_b$ list be empty. Let the costs of all the other nodes be infinity.
3. Alternate the forward and backward search directions based on a static or dynamic criterion. If there are no nodes on the OPEN list for the selected direction, try the opposite-direction list. If both OPEN lists are empty, report failure. Otherwise, select the node on the OPEN list with the lowest cost ($f'$ value) and call it BEST. Remove this node from the OPEN list and place it on the CLOSED list. See if BEST is a node that satisfies a stopping criterion. If so, go to step 4. Otherwise, generate the successors of BEST based on its connecting segment attributes contained in the map database. For each such successor node $n$, perform the following steps:
    3a. Compute the cost of $n$: $g(n)$ = cost of BEST + cost from BEST to $n$.
    3b. If $n$ matches a node already in OPEN, check to see if node $n$ has a lower cost ($g$ value). If so, replace the matching node's cost with the cost of node $n$, and set the matching node's back pointer to BEST.
    3c. If $n$ matches a node already in CLOSED, check to see if node $n$ has a lower cost ($g$ value). If so, replace the matching node's cost with the cost of node $n$, set the matching node's back pointer to BEST, and move the matching node to OPEN.
    3d. If $n$ is not already in either OPEN or CLOSED, set node $n$'s back pointer to BEST and place node $n$ in OPEN. Repeat step 3.
4. From BEST, traverse the forward back pointer up to the origin node, the backward pointer up to the destination node, link these lists together and report the solution route.

Recall that the running time for the unidirectional search (modified shortest path algorithm) is $O(b^d)$. The bidirectional search has running time $O(2b^{d/2}) = O(b^{d/2})$. From this calculation, we see that a bidirectional search will explore fewer nodes than a unidirectional search, as shown in Figure 5.7. This leads to a fast computation time for the bidirectional search, verified experimentally in [15].

Similarly, a bidirectional heuristic search can also be used. In Figure 5.8 the abstract spaces explored by the unidirectional and bidirectional heuristic search algorithms are shown.

If we search in both directions, we have the bidirectional heuristic search algorithm listed below. As in Section 5.3, the difference is highlighted by a vertical bar in step 3d.

### Bidirectional Heuristic Search Algorithm

1. Let the initial forward $OPEN_f$ list contain only the origin node with zero cost ($g$ value) and the forward $CLOSED_f$ list be empty. Let the costs of all other nodes be infinity.
2. Let the initial backward $OPEN_b$ list contain only the destination node with zero cost ($g$ value) and the backward $CLOSED_b$ list be empty. Let the costs of all the other nodes be infinity.
3. Alternate the forward and backward search directions based on a static or dynamic criterion. If there are no nodes on the OPEN list for the selected direction, try the opposite-direction list. If both OPEN lists are empty, report failure. Otherwise, select the node on the OPEN list with the lowest cost ($f'$ value) and call it BEST. Remove this node from the OPEN list and place it on the CLOSED list. See if BEST is a node that satisfies a stopping criterion. If so, go to step 4. Otherwise, generate the successors of BEST based on its connecting segment attributes contained in the map database. For each such successor node $n$, perform the following steps:
3a. Compute the cost of $n$: $g(n)$ = cost of BEST + cost from BEST to $n$.
3b. If $n$ matches a node already in OPEN, check to see if node $n$ has a lower cost ($g$ value). If so, replace the matching node's cost with the cost of node $n$, and set the matching node's back pointer to BEST.
3c. If $n$ matches a node already in CLOSED, check to see if node $n$ has a lower cost ($g$ value). If so, replace the matching node's cost with the cost

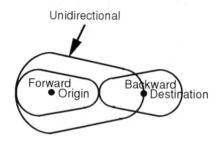

**Figure 5.8** Spaces examined by unidirectional and bidirectional heuristic search algorithms.

of node *n*, set the matching node's back pointer to BEST, and move the matching node to OPEN.

    3d. If *n* is not already in either OPEN or CLOSED, set node *n*'s back pointer to BEST, place node *n* in OPEN, and compute the heuristic evaluation function (cost) for node *n*: $f'(n) = g(n) + h'(n)$. Repeat step 3.

  4. From BEST, traverse the forward back pointer up to the origin node, the backward pointer up to the destination node, link these lists together and report the solution route.

Ideally, the bidirectional search should meet at the middle (Figures 5.7 and 5.8). As discussed in [15], if there are several different paths from the origin to the destination, the search may not meet in the middle, as shown in Figure 5.9. The two searches may each find nearly complete paths before intersection occurs, or one may find a complete path and the other almost a complete path before they intersect. In other words, there may be duplication of effort. In the worst case, such as an improper termination criterion for a heuristic evaluation function that considers only the estimated travel cost, a bidirectional heuristic search could double the work of a unidirectional heuristic search. Therefore, the choices of termination criterion and heuristic evaluation function are very important for a proper implementation of a bidirectional heuristic search.

As discussed in the section on the heuristic search, the A* algorithm is capable of finding an optimal route if the admissibility property is satisfied. A bidirectional heuristic search must explore almost twice as many nodes as a unidirectional heuristic search to determine the optimal route. The reason is that a bidirectional search must compare all the connected routes during the search to ensure the route is optimal. In real life, human beings seldom need to search for an optimal solution during their daily activities. A near-optimal solution is often a satisfactory solution for many problems. If an optimal solution is not required, the bidirectional search is a very good candidate for route planning. In other words, the optimality constraint should be relaxed when using a bidirectional search algorithm. By relaxing the optimality

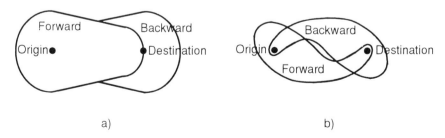

                 a)                                              b)

**Figure 5.9** Spaces examined by bidirectional heuristic search algorithms with (a) a poor termination criterion and (b) a poor heuristic evaluation function.

constraint, the bidirectional search will find a near-optimal route long before the optimal solution is reported.

There are several different variations of the bidirectional search algorithm that can be used to speed up the search process. In the algorithms discussed earlier, the travel cost from the origin or destination is used to determine which routes (nodes) need to be further expanded. One method involves instead the use of the number of nodes expanded to make such decisions. This method results in selecting routes generated from the origin with about the same number of nodes as the routes generated from the destination. Another method involves using a weighting factor for each road segment during the search, such that a higher weighting factor is used for segments close to the straight line between origin and destination than for segments further away from this line. The algorithm thereby forces the search process to select a route that is close to the straight line connecting the two starting nodes. Yet another variation is to use road type or class to reduce the amount of unnecessary searching. This operates on the assumption that every road needs to be examined in the areas near the origin and destination, but beyond these areas only major roads need be considered in the search. Clearly, its operation reduces the total number of nodes examined. This leads naturally to our next topic of discussion.

## 5.5  HIERARCHICAL SEARCH

There is a natural hierarchy of roads, with expressways at the top and side streets at the bottom. This leads to the idea of a more efficient, hierarchical search algorithm for a given road network. The basic idea is to first perform a search in an *abstraction space* rather than in the entire original problem space. An abstraction space is a simplified representation of the problem space in which unimportant details are ignored. The search then fills in the details from the original space based on the results obtained in the search on the abstraction space. Choosing a simplified representation of this type leads to a significant increase in problem-solving ability and efficiency.

Human beings are well aware of the value of abstraction. For instance, when solving a very complex problem, we often first ignore the low-level details and concentrate on the essential features of the problem. We then fill in the details. This idea is easily generalized to multiple hierarchical layers of abstraction such that each layer contains a different level of detail. Because the computer was developed to help human beings, it is very natural for the problem of abstraction to have been addressed in the field of computer science, especially artificial intelligence.

The main reason that abstraction reduces the complexity of problems is that the total complexity becomes the sum of the complexities of the individual searches, rather than the product of the complexities [23]. This method has been verified experimentally to be very effective in reducing the search space for complex problems [24]. The problem of abstraction was analyzed in detail and formalized in [25]. To make analysis of the hierarchical search easier, first assume that the running time for a given search algorithm is $O(b^d)$. For a single layer of abstraction, the hierarchical search can reduce a search algorithm with running time $O(b^d)$ to $O(b^{d/2})$. For multiple layers of abstraction, the optimum abstraction hierarchy for a problem space with $b^d$ states (in our case nodes) will have $\ln b^d$ layers of abstraction. The size ratio for successive layers of abstraction is $e$. This type of optimum abstraction hierarchy can reduce the search time from $O(b^d)$ to $O(d \log b)$. In other words, it has been found [25] that the optimum hierarchy has a logarithmic number of layers with a constant size ratio, and that this hierarchy reduces the search from being exponential in the number of states in the problem space to being linear. Certain search algorithms have a performance problem in that the number of nodes to be expanded in the road network often grows exponentially. From this analysis, we see that multiple layers of abstraction can reduce exponential complexity to linear complexity. It is therefore clear that a hierarchical search can potentially reduce the time complexity of a search algorithm.

To implement a hierarchical search and to simplify the route-planning task, the map database needs to be constructed hierarchically as discussed in Chapter 2. Others have also proposed and implemented hierarchical map databases to aid route planning [16,26,27]. The higher a hierarchical database layer, the less detailed the abstracted representation. Because a higher layer abstraction space contains fewer nodes and segments than its underlying search spaces, a search in an abstract space often yields an incomplete solution (but with reduced search time). The nodes in the abstract space can be either regions in the original space or a subset of the original nodes. Clearly, during the search we need to switch between different layers to take advantage of the hierarchy. Avoiding overhead during layer switching and determining when to switch to a different layer play important roles in the algorithm efficiency.

In practice, it is very difficult to construct a database or to design an algorithm that will satisfy the conditions of optimum abstraction hierarchy as discussed. For vehicle navigation, it is natural to use road ranks to define a hierarchy for route planning. Despite the difficulty of achieving the theoretically optimum hierarchy, the bidirectional heuristic-based hierarchical search is still the most popular algorithm for route planning in a very large map database.

## 5.6 OTHER ALGORITHMS

Route planning can also be performed using a divide and conquer method [28,29]. The basic idea is to divide the road network into regions (blocks) and precompute optimal regional routes for each region and then find and assemble these regional routes into a complete route from origin to destination. In addition to precomputation and prestorage of regional routes in the first stage, the algorithm must be coupled with the ability to determine entry and exit nodes for each route in each region and sequentially connect them into a final route solution in the second stage. Because all optimal routes within a region must be prestored in memory, Dijkstra's algorithm (unmodified) is generally used for precalculation. A different algorithm may be used for the second stage. Depending on how the road network is divided into regions and how the information available in the precomputed regional routes is used, various techniques can be used to provide the final solution to the vehicle user. This method can also be viewed as an approach to reconstruct the road network hierarchically (two layers in the preceding discussion) and then precompute all origin and destination pairs in the network. A similar idea can be extended to multivehicle route planning in hierarchically structured multilayered databases [30].

The real-time heuristic search algorithm proposed in [31] can be modified for route planning in certain situations [16], which can be classified as an anytime algorithm. Unlike the A* algorithm, which finds the optimal route, this algorithm finds a satisfactory route. For a real-time embedded in-vehicle navigation system and a very large map database, time constraints usually place a limit on the number of nodes that can be searched. For example, a sequential computer may need to switch to a more critical task before the entire search is completed. This fact can be represented by a *search horizon*. In a real-time heuristic search algorithm the search horizon is a search depth determined by the time available for searching the road network (or a directed graph tree during the search). This limited search horizon causes the real-time heuristic search to make a decision based on local optimality. The depth of the search forward from the current node (state) is determined by the computer resources available, that is, information provided via a recent update from the dynamic map database (containing updated dynamic traffic information), CPU time, or the time available for planning. If a more powerful CPU becomes available, the search depth can easily be increased, that is, the search horizon can be adjusted according to the available hardware and software. By contrast, neither the heuristic search algorithm nor the search algorithms discussed in this chapter place any restrictions on the search horizon to limit the use of either CPU or memory. This could lead to problems involving excessive searches. For instance, a search process may fail to report a solution route due to resource limitations, even though at least one solution route does exist.

Algorithms developed for the traveling-salesman problem can be used to plan a trip with multiple stops [32]. This study has been generalized as the vehicle routing problem [33]. Although there are a number of extensions, the basic problem involves a vehicle to be stopped at various intermediate destinations before reaching a final destination during a journey or a list of customers to be served by a vehicle or a fleet of vehicles from a single warehouse. For some service problems, each customer may place a demand for a certain quantity of goods. There are limits on vehicle capacity, on the maximum distance that any particular vehicle can travel, etc. In short, the objective of the algorithm is to find a set of routes that minimizes the total travel cost. The vehicle routing problem is known as *NP-complete*, that is, it is unlikely that any polynomial-time algorithm exists for solving the problem. Historically, the focus has been on methods of generating a feasible solution which can then be improved using a local search. Genetic algorithms, tabu searches, and robust algorithms have recently become popular. Some of these algorithms have been expanded to cover multivehicle routing problems. For information on current trends, benchmark problems, and more references on the vehicle routing problem, see [34,35].

In recent years, interest in dynamic on-line algorithms for shortest path problems has grown. The main application domains for these algorithms involve dynamically maintaining maximum flow in a network, bipartite matching in graphs, incremental computations for data flow analysis, and interactive systems design. More information about these algorithms can be found in [36] and the references contained therein. These algorithms can be modified for use in vehicle route planning. The main goal of these algorithms is to design efficient data structures that offer fast solutions and can operate in dynamic environments where, for example, the input data change. All of the other search algorithms discussed in detail here are only capable of dealing with static input data. Given the capability of these dynamic algorithms, it should not be very difficult to adapt them for dynamic vehicle navigation under conditions where the road conditions (travel costs of road segments) are dynamically updated on the vehicle by either a traffic information center or a traffic management center.

We have now reached the end of this chapter. Before reading the next chapter, consider the following interesting questions. Assume that a bidirectional search algorithm is used in dynamic route planning. Suddenly, the travel cost of a two-way road is updated in one direction only while the route is being planned. Can this type of search still find a nearly optimum solution in reasonable time and space for any road network? On the other hand, can a dynamic algorithm do a better job planning the route in this situation?

## References

[1] T. H. Cormen, C. E. Leiserson, and R. L. Rivest, *Introduction to Algorithms*, Cambridge, MA: MIT Press; New York: McGraw-Hill, 1990.

[2] F. S. Hillier and G. J. Lieberman, *Introduction to Operations Research*, 6th ed., New York: McGraw-Hill, 1995.

[3] H. A. Taha, *Operations Research: An Introduction*, 5th ed., New York: Macmillan, 1992.

[4] M. Kantrowitz, *CMU Artificial Intelligence Repository*, http://www.cs.cmu.edu/ Web/Groups/AI/ html/repository.html, 1996.

[5] E. W. Dijkstra, "A Note on Two Problems in Connexion with Graphs," *Numerische Mathematik*, Vol. 1, 1959, pp. 269–271.

[6] E. F. Moore, "The Shortest Path Through a Maze," *Proc. International Symposium on the Theory of Switching*, Cambridge, MA: Harvard University Press, Apr. 1959, pp. 285–292.

[7] A. V. Aho, J. E. Hopcroft, and J. D. Ullman, *Data Structures and Algorithms*, Reading, MA: Addison-Wesley, 1983.

[8] R. K. Ahuja, K. Mehlhorn, J. B. Orlin, and R. E. Tarjan, "Fast Algorithms for the Shortest Path Problem," *J. Assoc. Computing Machinery*, Vol. 37, No. 2, Apr. 1990, pp. 213–223.

[9] A. Barr and E. A. Feigenbaum (Eds.), *The Handbook of Artificial Intelligence*, Los Altos, CA: William Kaufmann, 1981.

[10] Y. Zhao, "Adaptive Real-Time Route Planning for IVHS," *Fall 1989 Summary Report*, IVHS Technical Report No. 89-A, The University of Michigan, Dec. 1989.

[11] R. J. Russell and P. Norvig, *Artificial Intelligence: A Modern Approach*, Englewood Cliffs, NJ: Prentice Hall, 1995.

[12] P. E. Hart, N. J. Nilsson, and B. Raphael, "A Formal Basis for the Heuristic Determination of Minimum Cost Paths," *IEEE Trans. Syst. Sci. Cybernetics*, Vol. 4, No. 2, July 1968, pp. 100–107.

[13] Y. Zhao, "Theoretical and Experimental Evaluation of a Local-Minimum-Recovery Navigation Algorithm," in *Recent Trends in Mobile Robots*, Yuan F. Zheng (Ed.), Singapore/New Jersey: World Scientific, 1993, pp. 75–117.

[14] E. Rich and K. Knight, *Artificial Intelligence*, 2nd ed., New York: McGraw-Hill, 1991.

[15] I. S. Pohl, "Bidirectional Search," in *Machine Intelligence*, B. Meltzer and D. Michie (Eds.), New York: American Elsevier Publishing Co., 1971, pp. 127–140.

[16] Y. Zhao and T. E. Weymouth, "An Adaptive Route-Guidance Algorithm for Intelligent Vehicle-Highway Systems," *Proc. American Control Conference*, June 1991, pp. 2568–2573.

[17] P. C. Nelson, C. Lain, and J. Dillenburg, "Vehicle-Based Route Planning in Advanced Traveler Information Systems," *Proc. Intelligent Vehicles Symposium*, July 1993, pp. 152–156.

[18] L. R. Rilett, C. Blumentritt, and L. Fu, "Minimum Path Algorithms for In-Vehicle Route Guidance Systems," *Proc. IVHS America 1994 Annual Meeting*, Apr. 1994, pp. 27–36.

[19] S. Shekhar and A. Fetterer, "Path Computation in Advanced Traveler Information Systems," *Proc. 1996 Annual Meeting of ITS America*, Apr. 1996, pp. 304–312.

[20] G. B. Dantzig, *Linear Programming and Extensions*, Princeton, NJ: Princeton University Press, 1963.

[21] T. A. J. Nicholson, "Finding the Shortest Route Between Two Points in a Network," *The Computer Journal*, Vol. 9, No. 3, Nov. 1966, pp. 275–280.

[22] I. S. Pohl, "Bidirectional and Heuristic Search in Path Problems," Ph.D. dissertation, Ann Arbor, MI: University Microfilms International, 1969.

[23] M. Minsky, "Steps Toward Artificial Intelligence," in *Computers and Thought*, E. A. Feigenbaum and J. Feldman (Eds.), New York: McGraw-Hill, 1963, pp. 406–450.

[24] E. D. Sacerdoti, "Planning in a Hierarchy of Abstraction Space," *Artificial Intelligence*, Vol. 5, 1974, pp. 115–135.

[25] R. E. Korf, "Planning as Search: A Quantitative Approach," *Artificial Intelligence*, Vol. 33, 1987, pp. 65–88.

[26] T. Hashimoto, K. Hirano, and Y. Kobayashi, "Development of Route Calculation System," presented and distributed at *Intelligent Vehicles Symposium*, June/July 1992.

[27] T. Yagyu, M. Fushimi, Y. Ueyama, and S. Azuma, "Quick Route-Finding Algorithm," *SAE Paper No. 930555*, Society of Automotive Engineers, 1993, pp. 121–126.

[28] C. B. Harris, R. Goss, E. J. Krakiwsky, and H. A. Karimi, "Optimal Route Information System—Automatic Vehicle Location (AVL) Route Determination," *Proc. Urban Regional Information Systems Association (URISA)*, Nov./Dec. 1987, pp. 29–41.

[29] M. Sugie, O. Menzilcioglu, and H. T. Kung, "CARGuide—On-Board Computer for Automobile Route Guidance," *Proc. 1984 National Computer Conference*, July 1984, pp. 695–706.

[30] Y.-H. Huang, E. A. Rundensteiner, and N. Jing, "Evaluation of Hierarchical Path Finding Technologies for ITS Route Guidance," *Proc. 1996 Annual Meeting of ITS America*, Apr. 1996, pp. 340–350.

[31] R. E. Korf, "Real-Time Heuristic Search," *Artificial Intelligence*, Vol. 42, Mar. 1990, pp. 189–211.

[32] G. Reinelt, *The Traveling Salesman: Computational Solutions for TSP Applications*, Berlin: Springer-Verlag, 1994.

[33] N. Christofides, "Vehicle Routing," in *The Traveling Salesman Problem: A Guided Tour of Combinatorial Optimization*, E. L. Lawler, J. K. Lenstra, A. H. G. Rinnooy Kan, and D. Shmoys (Eds.), New York: Wiley, 1985.

[34] D. J. Bertsimas and D. Simchi-Levi, "A New Generation of Vehicle Routing Research: Robust Algorithms, Addressing Uncertainty," *Operations Res.*, Vol. 44, No. 2, Mar./Apr. 1996, pp. 286–304.

[35] T. Duncan, *The Vehicle Routing Problem*, http://www.aiai.ed.ac.uk/ ~timd/vehicles/vrp.html, 1996.

[36] H. N. Djidjev, G. E. Pantziou, and C. D. Zaroliagis, "On-Line and Dynamic Algorithms for Shortest Path Problems," *STACS 95: 12th Annual Symposium on Theoretical Aspects of Computer Science (Proceedings)*, Berlin: Springer-Verlag, March 1995, pp. 193–204.

# CHAPTER 6
▼▼▼

# ROUTE GUIDANCE MODULE

## 6.1 INTRODUCTION

Route guidance is the process of guiding the driver along the route generated by the route planning module. Guidance can be given either before the trip or in real time while en route. The pretrip guidance could be presented to a driver as a printout. Such a printout would be similar to travel tips provided by a travel agency, but with detailed turn-by-turn driving instructions, or just like the output generated at a kiosk near the car rental counter. These instructions might include turns, street names, travel distances, and landmarks. On the other hand, en-route guidance would require providing turn-by-turn driving instructions to a driver in real time. It is much more useful, but requires a navigable map database, an accurate positioning module, and demanding real-time software and computation power. In this chapter, we concentrate on real-time route guidance.

The route guidance module uses the outputs of both the route-planning module and the positioning subsystem to guide the vehicle on the road. In addition to a map database module, the positioning subsystem could consist of the positioning module alone or a positioning module and a map-matching module. Once a particular route has been generated by the route-planning module and the position of the vehicle has been determined by the positioning subsystem, the route guidance module needs to coordinate with these subsystems to present proper guidance to the driver.

---

The presentation is aided by the human-machine interface module to be discussed in the next chapter. A simplified functional diagram of the interaction between the route guidance module and the other modules is shown in Figure 6.1. All modules except for the human-machine interface module have already been discussed in previous chapters. Note that in some systems the maneuver generation task is an integral part of the route-planning module. We discuss it here primarily for purposes of exposition.

As the vehicle moves, real-time route guidance requires that the location of the vehicle as a function of time continually be compared to the best route generated by the route-planning module. The real-time route guidance system continuously updates its knowledge of the vehicle's current position and direction of travel as well as the road on which the vehicle is traveling. As a turn or a maneuver approaches, the route guidance system alerts the driver with visual signals, audible signals, or driving instructions.

## 6.2  GUIDANCE WHILE EN ROUTE

As indicated in Figure 6.1, route guidance consists of two tasks, a maneuver generation task and a route-following task. Note that there are many different designs and implementations for the route guidance module other than the one sketched in Figure 6.1. For instance, the maneuver generation task could be an integral part of the route-planning module. Both the maneuver generation task and route-following task might deal with the data generated by the route-planning module directly. To

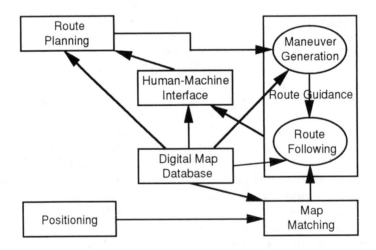

**Figure 6.1** Simplified diagram showing the interaction between the route guidance module and the other modules.

convey the basic ideas, however, we use this simplified functional diagram to describe the principles and practical issues involved in this module. Many of the approaches described in this section are applicable to the other situations discussed later in this chapter.

### 6.2.1 Maneuver Generation

As mentioned, the maneuver generation task can be either a by-product of the route-planning module or an individual task within the route guidance module. The main function of this task is to generate a maneuver list that can be followed by the route-following task. The typical output route generated by a nonhierarchical route-planning module is a simple list of road segments that does not provide enough guidance information. The maneuver generation task can then augment each road segment in the list with appropriate maneuver and turn type information that can later be interpreted by the human-machine interface in the form of specific visible and audible instructions supplied with assistance from the route-following task. Figure 6.2 lists some samples of these types. Different combinations of maneuver and turn types may correspond to different visual and audible presentations to the driver.

If a hierarchical route plan module is being used, it may be necessary to break the output segments into the lowest layer segments before augmenting them as discussed earlier, depending on how high-layer segments are constructed. Recall that in Section 2.6 we discussed the fact that in some hierarchical map databases the

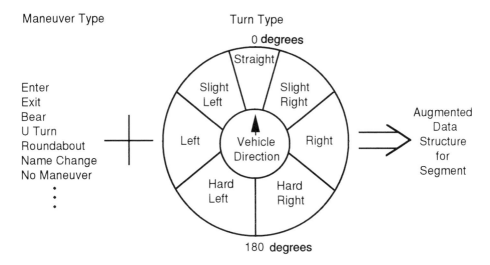

**Figure 6.2** Maneuver and turn types used for augmentation of the segment data structure.

segments at higher layers are equivalent to a combination of several segments stored in the layer immediately below. To save memory, the segments in the higher layers only contain information relevant to route planning. For positioning and guidance, it is better to use the segments in the lowest layer, which contains detailed guidance information on each segment. Also, almost all vehicles use segments at both high and low levels during any given trip, which means that it is much more efficient to guide the vehicle using the lowest layer where all relevant data are stored. In fact, the data access and storage methods used during planning and guidance are critical to achieving maximum performance, particularly in a system based on a hierarchical map database.

Careful consideration should be given to memory space and data access mechanisms in order to guarantee efficient implementation of the route guidance module. The higher layer data may be in a form suitable only for planning to minimize the memory usage of the route-planning module. As we saw in Section 5.5, route planning will benefit considerably from the use of a hierarchical data representation for the map. One side effect is that extra time may be required to process the resulting route in the route guidance module. Of course, there will always be a trade-off. For a system designed to travel on a large road network, it is usually (based on current technologies) worth the trouble to do a little more work in guidance, compared with the time saved in planning. Furthermore, currently available real-time computing technology allows multiple tasks to be executed simultaneously. Once just enough maneuvers to enable initial route following have been computed, the remainder of the work can be performed in background. After all, the on-board navigation system is a real-time embedded computer application. It is much easier in terms of keeping the map-matching-based positioning simple to report the current vehicle position based on the basic layer (lowest one). In other words, a route planned on the basis of certain hierarchical databases may not be sufficiently detailed for guidance. A maneuver list based on the basic segments with sufficient detail for guidance should be generated. In a hierarchical implementation, the maneuver generation task might include breaking all higher layer segments on a route into basic road segments while adding maneuver and turn type information to the maneuver list.

If a slow secondary or remote storage device is used for the hierarchical map database, efficient use of memory space requires that we keep at least the information required for route planning in main memory so that only the information on the relevant road segments needs to be retrieved. In other words, the information required for positioning, maneuver generation, and route following should be kept separate from that needed for route planning in order to maximize planning efficiency when the database is stored on a slow device. After route planning has been completed, we need to maintain detailed information in main memory only for those route segments on the maneuver list and in the immediate vicinity of the current vehicle location. The memory required for route storage should appear similar to the body of a snake that has recently swallowed an egg, with the body around the egg representing route storage space for the map area in the immediate vicinity of the vehicle, and the rest of the body representing the route storage space for areas that are not near the current location of the vehicle. Similar techniques have been used in mobile robot guidance [1]. To save memory space, we can also discard any data used by the route-planning module that are not necessary for route guidance.

Good data structures should be used for efficiency of data access [2,3]. These data structures should be able to help the system avoid exhaustive searches, on-demand calculations, and repeated accesses of the same data. For instance, a hash table can be used for quick and frequent look-up of the current location of the vehicle. Ideally, once one segment on the route is found, identifying the upcoming segment should be as simple as accessing a structure field or following a pointer. One possible data structure of this type is a double linked list of maneuvers with a back pointer to the last maneuver just accessed and a next pointer to the next maneuver. Enough calculated travel cost results can be stored to limit on-demand computation during route guidance. If there is insufficient main memory available to share with other navigation tasks, a *cache* may be needed to manage data access and avoid repeated requests for the same data from secondary memory. A cache is a collection of data blocks that are logically part of secondary memory, but which are kept in the main memory for performance reasons. In other words, a cache provides a high-speed buffer in main memory to act as a temporary storage area for the secondary memory. In summary, we should store as little guidance data as possible in main memory while still providing fast access and a good balance between time and space tradeoffs.

## 6.2.2  Route Following

On-road guidance is not difficult, provided the current position of the vehicle can be determined precisely. Once the position is known, proper signals and instructions can be prepared for or presented to the driver after comparison of the position against the planned route. This planned route consists of a sequence of road segments that leads to a destination generated under planning conditions determined either by the system design or by driver-selected planning criteria. At any point during the trip, the vehicle should be traveling on a segment contained in the planned route. The route-following task closely monitors the position of the vehicle on this segment to determine when to take proper guidance action. These actions are presented to the driver through the human-machine interface. In short, the main function of the route guidance module should be to decide when and how to guide the driver.

Once the route-following task has identified the driver's next maneuver, it must have a way to convey this information to the driver. The most popular method in current navigation systems is to provide one or a series of voice announcements to warn the driver of the approaching maneuver so that he or she can take appropriate action (see Figure 6.3). Other methods could involve displaying the maneuver instructions on a screen either manually or automatically, providing simple audio tones, etc. The announcement can be made through a speaker using either a prerecorded human voice or a synthesized voice as discussed in Section 7.3.

One study recommends using certain message types, contents, and timings when a series of three messages is issued for each maneuver [4]. These three messages are called "early," "prepare," and "approaching." The "early" message warns the driver of an upcoming maneuver (this message should always be given). The "prepare" message informs the driver to move to the appropriate lane and begin searching for the street, landmark, or exit (this message may be omitted if there is no time to announce it). The "approaching" message instructs the driver to perform the maneuver (this message is a final reminder and should always be given). The content and timing of each message are listed in Table 6.1 [4]. To facilitate study and compare with other approaches, we have rounded off the distances in meters and provided approximate equivalents in miles.

Another study proposes the following timing for maneuver messages [5–7]:

- For nonexpressways (or urban areas), 300m before next maneuver and 700m if lane changes are necessary;
- For expressways (or rural areas), 500m before entrances and 2,000m before junctions and exits.

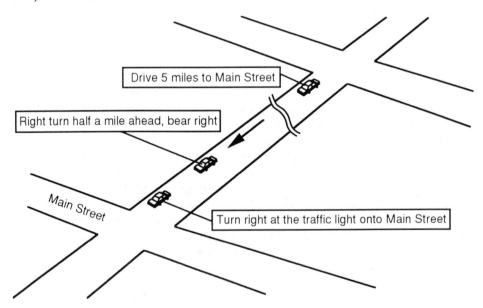

**Figure 6.3** Maneuver announcement during the journey.

**Table 6.1**
Content and Timing for Maneuver Messages

| Message | Content | City Timing | Highway Timing |
|---|---|---|---|
| Early | "In" {distance} "at" {location} {action} | 5 sec after turn | 15 sec after turn |
| Prepare | {distance} {location} {action} | 1500m | 3000m |
|  | or | (0.932 miles) | (1.865 miles) |
|  | {distance} {landmark} {location} {action} | Before turn | Before turn |
| Approaching | "Approaching" {location} {action} | 150m | 500m |
|  | or | (0.093 miles) | (0.311 miles) |
|  | "Approaching" {landmark} {location} {action} | Before turn | Before turn |

If two maneuvers are too close to one another, messages for both intersections will be given before the first intersection [8]. If the next maneuver follows the turn by a large distance, an "early" message will be given after the turn to warn the driver. If the driver presses the Verbal Guidance switch and a message containing turning information for the next maneuver is announced. Note that the two proposals discussed here consider only two speeds, that is, city roads and highway roads. This may not be sufficient for a broader range of road types.

Many of these message announcement (maneuver) timing proposals are based on experience gained from actual route guidance systems. From these proposals, we can see that these maneuver timings depend on the road type and distance to or after the maneuver. This implies that the road type, intersection type, travel speed and distance, and number of lanes are important factors to be considered when designing the maneuver timing of a route guidance module. Other reasoning techniques (such as a fuzzy-logic-based approach [9]) may also be used to combine all relevant factors and determine the best timing. Furthermore, the exact timing of advance messages such as the "early" and "prepare" messages is not critical, but the timing of turn messages such as the "approaching" message is. If the maneuver instructions are stored in a slow secondary memory device, a pre-fetching technique or cache may be required to make them available as an anticipated maneuver approaches. Whichever approach is used, the main purpose of this subsystem is to avoid announcing upcoming maneuvers either too early or too late. In this way, the driver can benefit from the instructions to increase safety and achieve maximum performance.

## 6.3 GUIDANCE WHILE OFF-ROUTE

Drivers may occasionally drive off the planned route for various reasons. They may make a wrong turn at an instructed intersection, may be unable to change to the correct lane for an intersection, may temporarily get off the route, etc. A mechanism to get them back on route must exist once the system has determined that the vehicle is no longer on the route (or not traveling on any road segment within a predetermined vicinity).

Different methods can be used to solve this problem. A simple solution is to alert the driver that the vehicle is off its route once this scenario is confirmed. A directional arrow then displays the direction toward the original destination on the screen. The driver should use these rough directions to get back to the original road or find a way to reach the destination. Another solution is to notify the driver that the vehicle has gone off-route and that it needs to replan a new route to the original destination. Replanning a short route from the current off-route location to the original route is another option. The next on-route segment following the segment where the vehicle went off-route could be used as an intermediate destination in this circumstance. One could also replan a route using the original destination. The tricky part of this solution is how to determine a new origin for the route-planning system. If the vehicle is stopped after the off-route occurred, the current location of the vehicle can be used as the origin. Otherwise, we must select an origin that is ahead of the current travel direction of the vehicle. One good candidate is the node that will be reached in a travel time equal to the maximum amount of time assumed necessary to replan the new route.

All of the methods discussed assume that the vehicle is off the planned route but still on road. How do we guide an off-road vehicle? Clearly, an off-road vehicle is off the planned route too. One strategy is to notify the driver that the vehicle is off the road and that it needs to be driven back to the nearest road. The system keeps monitoring to see whether the vehicle has returned to the road network. Once it determines that the vehicle is on the road again, the preceding methods can be used to guide the vehicle again. We can also try to direct the driver back to the nearest road by pointing in that direction on the screen or determining and displaying the nearest road segment to which the driver should return. There are many other approaches in addition to those mentioned here that can also be used as off-road recovery mechanisms.

## 6.4　GUIDANCE WITH DYNAMIC INFORMATION

In the previous discussion, we implicitly assumed that the route guidance module was based on a static route-planning module. A system based on static route planning uses the static travel cost of each road segment to calculate the minimum-cost route and to guide the vehicle along this route. As mentioned in Chapter 5, this cost may involve distance, travel time, minimum number of turns, etc. This static cost may be a value derived from a prestored map database that is independent of the time of day or a statistical value that has been predetermined for this particular segment during a certain time period. The static cost is an oversimplified assumption. In the real-world environment, the actual cost is a dynamic value that depends on traffic conditions at a specific time. Therefore, a system based on dynamic route planning is preferable from the point of improved route guidance. Ideally, the route used for guidance would be calculated on the basis of the real-time cost of each particular road segment. To be more specific, this dynamic travel cost should be based on the link cost, which is the segment cost plus the turning cost received during the trip from the traffic control center or traffic management center. Remember that the cost of left turns, right turns, turns at stop signs, and turns at traffic lights all depend on the specific road.

Similar to the route-planning module, dynamic route guidance (or dynamic navigation) can be classified into either multivehicle (system-wide) dynamic route guidance, which guides all vehicles on the road network while minimizing the total travel cost of all the vehicles, or single-vehicle dynamic route guidance, which guides one vehicle while minimizing the travel cost of this particular vehicle. We discuss only the single-vehicle dynamic route guidance case in this section.

Assume that the system in the vehicle is capable of receiving, filtering, and storing information received from a control center or information center (such as a traffic information center). The route guidance module must take appropriate actions in response to incoming congestion reports for the planned route.

In dynamic route guidance systems, we must face three important issues with regard to the system configuration: First, we need to determine whether the dynamic (and some of the static) information should be processed by the central system (host end) or the on-vehicle system (mobile end); second, we need to determine the dynamic information update frequency; and, third, we need to determine the communication medium to be used in ensuring smooth information flow between the infrastructure and vehicle. The communication medium is discussed in Chapter 8. In this section, we limit ourselves to a configuration in which the static travel information is stored on board and the dynamic information is provided periodically by a traffic control center or traffic management center. We assume that the information update rate occurs frequently enough to reflect the traffic conditions on the road and route planning is done in the vehicle. In this way, the guidance activities are the sole responsibility of the mobile vehicle but not the centralized host. In Chapter 10, this is classified as a distributed navigation approach. We now discuss one dynamic route guidance algorithm that could be used for this configuration (Figure 6.4). Note that some algorithms discussed in Section 5.6 can also be used. Similar approaches can be used for other configurations.

The main idea of this algorithm is to rely on the dynamic travel cost as much as possible for route guidance. To achieve this goal, the on-board navigation computer must be able to accept the dynamic cost data and consequently change the static database into a dynamic one. A simple method involves maintaining a

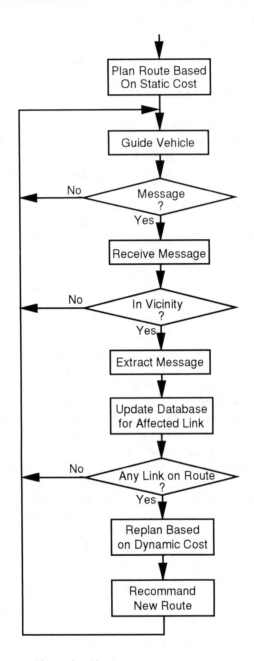

**Figure 6.4** A dynamic route guidance algorithm.

fixed static database on board. During the route guidance process, this database can be dynamically updated to reflect the new cost information.

Assume that the travel time is used as the route-planning cost. As we learned in Chapter 2, the length and speed limit of a road segment are attributes used in the static map database. These attributes can be easily converted into travel time for route-planning purposes as discussed in Chapter 5. With the travel time as the cost, the static route-planning system can then generate a minimum-travel-time route for the driver. This route is derived under the assumption that an average driver usually drives near the speed limit on every road segment. We know now that a better way of providing the travel cost information is to use a static database that more accurately reflects actual travel conditions because vehicles do not always travel at the allowed speed limit.

One example of this type involves integrating a historical profile with the database. This profile is originally obtained in an off-line process. Recall that each link represents a road segment plus information on turning delays. This process uses a detailed representation of a traffic flow model (supported by the road network and other data) to estimate link travel times and flows for origin-destination trip tables at each time of day [10]. The 24-hour day is divided into different periods so that the travel time for each link can be estimated appropriately. The model also incorporates traffic delay functions (which include turning movements and signal control parameters). When incorporated with the real implementation, these link travel times are validated and updated using actual data received from vehicles acting as traffic probes or from other traffic sensors and anecdotal information. Probe vehicles are vehicles that report back real-time traffic data for the roads on which they are traveling. This historical profile can be used as one of the inputs to the database compiler described in Section 2.6. After compilation, the link travel time stored in the profile becomes an attribute for the navigable database. When used in conjunction with a route guidance algorithm, it is hoped the system will present a realistic route. The historical profiles can be further improved at periodic intervals after system deployment on the basis of new traffic data, provided this feature has been included in the system. Despite this improvement over using the speed limit as the travel cost, the route recommended by the system will be still based on static costs prestored in a database. Only when a dynamic cost is used does the route guidance system become truly dynamic.

The algorithm described in Figure 6.2 first uses the static costs to plan a route. While the driver is following the planned route, the on-board system continually monitors the incoming message broadcast by a traffic control center or traffic management center. For simplicity, assume that the message contains travel link cost data for the area in which the vehicle is traveling. When the system receives a message, a filter is used to determine whether the message contains information for links in the vicinity of the vehicle. If the message contains such data, the affected links in the static database are updated using the new link cost information. The updated data can be placed in main memory (as shown in Figure 2.10 in Chapter 2) for fast storage and retrieval. This is not only convenient for maintenance of the dynamic database, but also beneficial for efficient route planning and guidance. If one or several of the affected links happen to fall on the portion of the preplanned route that the vehicle has not yet traversed, the route-planning module is automatically triggered to calculate (in the background) an alternative route based on these dynamic costs. If this alternative route is better than the current route, the driver is notified that a shorter route has been identified. A criterion for a route being better than the current route (such as saving of at least 20% in travel time relative to the current route) can be predefined. The driver can then accept or reject this recommendation. If the driver accepts this alternative route, the system will discard the current route and guide the driver along the new one.

Some of the implementation details for this dynamic route guidance algorithm have been omitted here. These include, but are not limited to, defining the incoming message frequency, defining what is meant by "in the vicinity of the vehicle," and defining the threshold for determining whether the travel time saved is worth alerting the driver about. The representation used for the road network and the dynamic model to be used are other areas that need to be addressed during implementation. Field validation and updating will be much easier if the representation of the road network and dynamic model are relatively accurate. Many representations and models have been proposed for estimating or predicting link travel times. These have been designed both for the normal situation and abnormal situations involving incidents or congestion. Various representations and models for the road network are described in [10–13]. See Chapter 11 for a working example of a dynamic route guidance system.

From common sense, dynamic route guidance should be able to direct drivers to less congested roads. Can you imagine what would happen if vehicles all equipped with the same dynamic guidance unit travel on a highly congested road network? If all these vehicles receive the same traffic information, they could be dispatched to the same, previously uncongested road. This road may very quickly become congested and make the traffic even worse. This phenomenon is known as the Braess paradox [14]. Clearly, this is the one situation that a traffic management system or a dynamic route guidance system must avoid. When this happens, it is not a good idea to direct all the vehicles to a single less-congested road.

The Braess paradox is caused by providing the same traffic information to all vehicles within a particular road network. Since dynamic route guidance has not been implemented on a widespread basis, most studies on this phenomenon are based on computer simulations. Some strategies have been proposed for dealing with this problem [14,15].

One strategy involves adding distortion factors to the link cost data. By doing this, we hope that different link cost data will be used by different vehicles in the area. From the last chapter, we know that when the costs are different, the route-planning algorithm will generally generate different routes. Clearly, guided by different routes the chance of congesting a single road is greatly reduced. However, if the link costs are distorted, we also know that the planning algorithm will not be able to find the optimal route in the absence of congestion. One issue still remains: selection of distortion factors for maximum efficiency of the road network. Another strategy involves broadcasting forecast link-cost data to the vehicles in the road network. Each vehicle reports its planned route to the traffic center. Based on the reported information, the center updates the forecast link-cost data and broadcasts it periodically. However, this requires each vehicle to have its own route planner. It also requires that the vehicle or infrastructure be able to provide enough communication bandwidth to send the information back to the center. Other strategies can also be used. More research as well as practical experience is needed in order to obtain enough data for further evaluation of these different strategies.

Before leaving this chapter, let us consider two questions. What is the best method for determining the new origin when replanning the route of a moving vehicle that has left the previously planned route? Besides the message announcement timing method introduced in this chapter, are there any other good methods for determining the best time to announce a maneuver instruction? As design engineers, we certainly do not want to surprise our drivers with any unexpected announcements, do we?

## References

[1] Y. Zhao, C. V. Ravishankar, and S. L. BeMent, "Coping with Limited On-Board Memory and Communication Bandwidth in Mobile-Robot Systems," *IEEE Trans. Syst. Man Cybernetics*, Vol. 24, No. 1, pp. 58–72, Jan. 1994.

[2] A. V. Aho, J. E. Hopcroft, and J. D. Ullman, *Data Structures and Algorithms*, Reading, MA: Addison-Wesley, 1983.

[3] T. H. Cormen, C. E. Leiserson, and R. L. Rivest, *Introduction to Algorithms*, Cambridge, MA: MIT Press; New York: McGraw-Hill, 1990.

[4] P. Green, W. Levison, G. Paelke, and C. Serafin, *Preliminary Human Factors Design Guidelines for Driver Information Systems*, Ann Arbor, MI: The University of Michigan, Transportation Research Institute, Technical Report No. UMTRI-93-21. Also published in McLean, VA: U.S. Dept. of Transportation, Federal Highway Administration, Report No. FHWA-RD-94-087, Dec. 1995.

[5] K. Kimura, K. Marunaka, and S. Sugiura, "Human Factors Considerations for Automotive Navigation Systems," *Proc. Triennial Congress of the International Ergonomics Association*, Vol. 4, Aug. 1994, pp. 162–165.

[6] H. Kishi and S. Sugiura, "Human Factors Considerations for Voice Route Guidance," *SAE Paper No. 930553*, Society of Automotive Engineers, 1993.

[7] E. J. Krakiwsky and R. L. French, "Japan in the Driver's Seat," *GPS World*, Oct. 1995, pp. 53–60.

[8] T. Ito, S. Azuma, and K. Sumiya, "Development of the New Navigation System—Voice Route Guidance," *SAE Paper No. 930554*, Society of Automotive Engineers 1993.

[9] Y. Zhao, A. M. Kirson, and L. G. Seymour, "Fuzzy Logic Applications in Vehicle Control and Navigation," *Abs. Third World Congress on Intelligent Transport Systems*, Paper #1225 (full paper in CD-ROM), Oct. 1996, p. 204.

[10] D. E. Boyce, A. Tarko, S. Berka, and Y. Zhang, "Estimation of Link Travel Times with a Large-Scale Network Flow Model for a Dynamic Route Guidance System," *Proc. IVHS America 1994 Annual Meeting*, Apr. 1994, pp. 37–43.

[11] J. F. Gilmore and N. Abe, "Neural Network Models for Traffic Control and Congestion Prediction," *IVHS Journal*, Vol. 2, No. 3, 1995, pp. 231–252.

[12] B. Ran, H. K. Lo, S. Weissenberger, and B. Hongola, "Predicting Dynamic Travel Times under Incidents in Transportation Networks," *Proc. IVHS America 1994 Annual Meeting*, Apr. 1994, pp. 37–43.

[13] V. Sethi, N. Bhandari, F. S. Koppelman, and J. L. Schofer, "Arterial Incident Detection Using Fixed Detector and Probe Vehicle Data," *Transportation Res. C*, Vol. 3, No. 3, Apr. 1995, pp. 99–112.

[14] L. R. Rilett and M. W. Van Aerde, "Modeling Distributed Real-Time Route Guidance Strategies in a Traffic Network that Exhibits the Braess Paradox," *Proc. IEEE-IEE Vehicle Navigation and Information Systems Conference (VNIS '91)*, Society of Automotive Engineers, Oct. 1991, pp. 577–587.

[15] M. Tokoro and S. Takaba, "Route Guidance Strategy Under High Penetration Condition," *Proc. IEEE Vehicle Navigation and Information Systems Conference (VNIS '94)*, Aug./Sep. 1994, pp. 303–308.

# CHAPTER 7
▼▼▼

# HUMAN-MACHINE INTERFACE MODULE

## 7.1 INTRODUCTION

The human-machine interface is a module that provides the user with the means to interact with the location and navigation computer and devices. This subject is often treated under ergonomics or human factors engineering in the literature. Researchers in the field of ergonomics conduct research and apply information about human behavior, abilities, limitations, and other characteristics to the design of tools, machines, systems, tasks, jobs, and environments [1]. The objectives are to increase productivity and safety, to maximize performance, and to ensure comfortable and effective human use. The reader may become familiar with various human factors problems and how to deal with them in general reference [2]. Information on automotive human factors engineering can be found in [3] and references contained therein. In addition to human factors for vehicle location and navigation systems, this chapter also addresses the basic principles of popular interface technologies.

To develop a successful human-machine interface, a certain procedure must be followed that may include identification of requirements, determination of functions to be supported, specification of interface type(s), selection of controls and displays, and, finally, designing and implementing these interfaces. We will concentrate on the various control devices and display technologies often used in vehicle location and navigation systems. These control devices and display technologies form an integral part of the human-machine interface module.

The primary design principles for a good human-machine interface module are as follows [4]:

1. The interface design should be consistent.
2. Controls and displays should function the way people expect them to function.
3. Controls and displays should be arranged in text-based order: from left to right, and from top to bottom.
4. The interface should minimize the need for the user to remember.
5. Operations that occur most often or have the greatest impact on driving safety should be the easiest to perform.
6. Controls, displays, and information elements that are used together should be adjacent to one another.

In short, a good design is often based on good engineering judgment and accepted human factors practice. It should stand the test of time and experience.

Every interface needs one or more control devices. Table 7.1 lists various control devices along with evaluation scores and recommended usage ([5], mostly from [6–8]). The evaluation scores are based on the following criteria:

- Usability under vibration conditions;
- Usability from multiple locations;
- Space required;
- Operability over desired temperature range;
- Any built-in display properties;
- Usability in noise;
- Susceptibility to dirt, grease, etc.;
- Hands-free use.

Whenever the control has one of the properties listed, it receives a score of 1. Otherwise, it receives a score of 0. The highest score should be 8. The vibration conditions described above mean those normally encountered inside an automobile traveling at highway speeds on a reasonably maintained concrete or asphalt roadway. The anticipated temperature range is approximately −34.4° to +54.4°C (−30° to +130°F).

From Table 7.1, we see that there is great variety of different computer and display control devices from which to choose. Although several devices have very similar scores, these scores result from different combinations of the criteria [5]. The key is to understand the advantages and disadvantages of each device and know when it should be used to achieve maximum performance while meeting proper safety guidelines. (For discussions of safety and driver performance under various

**Table 7.1**
Various Control Methods

| Control | Score | Recommended Usage |
|---|---|---|
| Foot push button | 6 | For use only when both hands occupied |
| Isometric joystick | 6 | For applications requiring return to center after each entry |
| Isotonic joystick | 6 | For controlling various display functions |
| Key-operated switch | 6 | For prevention of unauthorized machine operation and support of on/off functions |
| Keyboard | 4 | For use when vehicle is completely stopped (score for sealed keyboards) |
| Knob (continuous) | 5 | For use when low force or precise adjustment of a continuous variable is recorded |
| Legend switch | 6 | For display of qualitative information or other types of information when space is limited |
| Light pen | 2 | Use when vehicle is completely stopped and only for noncritical inputs |
| Push button | 5 or 6 | Support simple switching between two conditions |
| Rocker switch | 6 | As an alternative to toggle switches when two discrete positions are required |
| Rotary selector switch | 6 | For discrete functions when three or more detented positions are required |
| Slide switch | 6 | For two or more discrete positions |
| Toggle switch | 6 | For two discrete positions where space limitations are severe |
| Touch screen | 5 | For low training when targets are large, discrete, with less or no text input |
| Voice recognition | 6 | When the user's hands or eyes are busy |

human-machine interfaces, see Section 7.2.3, [9–11], and the other references listed at the end of this chapter.)

In the following sections, we first consider the commonly used interfaces based on visual displays. We then discuss several potentially useful voice-based interfaces.

## 7.2 VISUAL-DISPLAY-BASED INTERFACES

A visual display is an electronic device that converts electrical signals into a visual image in real time for direct interpretation by a human observer. It serves as the visual interface between human and machine. The display itself is typically an electro-optical device of some kind. It presents information within a fraction of a second from the time it was received, and continuously holds the information on the output device until new information is received. The image is created by making visible contrast patterns.

## 7.2.1 Display Technologies

For engineers, it is a challenging task to design or select a visual display for vehicle use because of the severe environment associated with such effects as varying light conditions, varying temperature, frequent vibration, and constant dirt and grease. In addition to these environmental constraints, we expect the display to be lightweight and have low power consumption. Because of the unique safety requirements, attention requirements, body multiplexing requirements, and visual environment in a vehicle, many human factors must also be considered.

Research suggests that the following functions can be supported using visual displays (modified from [12]):

- A wide range of quantitative and qualitative information can be displayed.
- Complex information can easily be extracted from the display.
- The display can emulate conventional displays such as paper maps.
- In-vehicle or heads-up displays become possible.

However, it is distracting if too much information is displayed or the information displayed is too complex. While driving, minimal information should be presented. This leads to the important subject of how to design the graphical user interface (GUI). Before discussing it, we first take a look at the design guidelines for the display interface used for GUI.

Many important factors need to be considered when selecting and using display technologies, such as display color, contrast, luminance, format, content, size, labeling, orientation, and placement. As discussed in [5] and the references contained therein, there should be no more than six symbol colors. Commonly accepted color conventions should be observed. Pure blue or pure red and pure blue displayed simultaneously on a dark background should be avoided. The user should be able to adjust the luminance level. The minimum monochrome contrast ratio (higher/lower) should be 1.5 for acceptable legibility and 2.5 for comfortable reading. The display should be positioned properly and the area surrounding the display screen should be matte black to avoid glare. The display luminance should range anywhere from 70 $cd/m^2$ to 150 $cd/m^2$, measured after any optical effects. It should be adjustable over a range of at least 50:1 (highest/lowest), preferably 100:1. The total number of operating hours until the display luminance is reduced to 50% of its original maximum luminance should be at least 5000 hr. The visibility envelope for the display should be no more than ±30 degrees to 40 degrees laterally from the design eye reference point and no more than 30 degrees above or 5 degrees below the design eye reference point. Interested readers may refer to [4,5] and the references contained therein for further discussion of other guidelines to help in the selection and evaluation of various technologies.

Based on the visual display characteristics and design guidelines just discussed, it becomes obvious that designing a safe and efficient GUI is another challenge for engineers. A variety of designs are possible, depending on the user specifications and the display devices used. These designs all have a very similar goal, that is, to simplify a variety of complicated functions so that the user will not be intimidated and can actually easily obtain useful information while driving. The interested reader is urged to refer to Chapter 11 and [4,13,14] for examples of actual and proposed user interfaces, as well as ITS-related magazines, conferences, and exhibitions to learn and see more about various different interfaces. A performance evaluation of different designs for touch screen keyboards can be found in [15], and the basic touch screen technologies are addressed in the next section. A more general discussion on GUI design can be found in [16].

Several different technologies are available for display devices. Table 7.2 lists various display devices, along with evaluation scores and recommended usage (mostly from [5]). The evaluation score is based on the following criteria:

- Salience in low ambient illumination;
- Salience in high ambient illumination;
- Salience in high ambient noise;
- Attention getting properties;
- Visibility from multiple locations;
- Usability if eyes are busy;
- Ability to display graphics;
- Ability to display same object in multiple colors;
- Space required;
- Operability over desired temperature range;
- Flexibility of formatting.

Whenever the display has one of the properties listed, it receives a score of 1. Otherwise, it receives a score of 0. The highest score should be 11. The anticipated temperature range is the same as in Table 7.1, that is, from approximately $-34.4°$ to $+54.4°C$ ($-30°$ to $+130°F$).

Early in-vehicle display screens generally used cathode-ray tube (CRT) displays. The inner screen of a CRT is coated with phosphors that emit light when struck by one or more beams of high-energy electrons. The electron beams are emitted from a cathode at the rear of the tube. On a raster-scanned CRT display, the electron beam sweeps across the entire display on each refresh cycle and illuminates the proper points as it comes to them. On a vector-scanned CRT display, the electron beam strikes only the points needed to be illuminated on the display. The color of the image is controlled by the selection of phosphors and electron guns.

An electroluminescent display (ELD) produces light from a polycrystalline phosphor on application of an electric field. The phosphors are sandwiched inside

**Table 7.2**
Various Display Methods

| Display | Score | Recommended Usage |
|---------|-------|-------------------|
| Analog/mechanical | 6 | For qualitative and quantitative value or qualitative value only |
| CRT | 7 | For text and graphics applications |
| Counter/mechanical | 6 | For presenting large ranges of quantitative data |
| ELD, matrix addressed | 6 | For text and graphics applications |
| HUD | 7 | To present an image overlaid on a view of an outside scene and to decrease visual transitions from "head-down" displays to the outside, forward scene |
| Indicator lights | 5 | To indicate system, equipment, and/or control condition |
| LED, indicator | 5 | For simple indications |
| LED, character based | 5 | For text applications |
| LED, matrix addressed | 5 | For text and graphics applications |
| LCDs | 4–6 | For text only or text and graphics applications. Use |
|    Reflective, character based | 5 | transflective LCDs when visibility in sunlight, low |
|    Reflective, matrix addressed | 5 | power consumption, and visibility of display from |
|    Transflective, character based | 6 | multiple viewer positions are more important than |
|    Transflective, matrix addressed | 6 | having a multicolor, high resolution display. Use |
|    Transmissive, character based | 4 | transmissive LCDs when display brightness, display resolution, and multicolor display are more |
|    Transmissive, matrix addressed | 6 | important than having visibility in sunlight, low power consumption, and visibility of display from multiple viewer positions. |
| PDP, character based | 4 | For text applications |
| PDP, matrix addressed | 6 | For text and graphics applications |
| VFD, character based | 6 | For text applications |
| VFD, matrix addressed | 8 | For text and graphics applications |

a grid of wires. Most ELDs are monochromatic, with a typical color being yellow on a black background. ELDs are often used as matrix-addressed text and graphics displays and as backlights for liquid crystal displays (LCDs). The main advantages of ELDs are ruggedness and ability to be viewed from any angle. The main disadvantages are higher production cost and limited multiple color capability.

A heads-up display (HUD) consists of an image source, projection optics, and a combiner. The image is generated in the image source and projected through collimation optics onto the combiner. It is viewed as an image overlaid on the forward visual scene. The forward windshield of the vehicle is often used for the HUD combiner. A vacuum fluorescent display (VFD) or a vector-scanned CRT is generally used as the image source.

A light-emitting diode (LED) produces light via injection of electrons into a solid-state semiconductor. An LED consists of a single $p$-$n$ junction, which is a semiconductor device. An LED emits light when the junction is forward biased so that minority carrier injection and electron-hole recombination occur. Different semiconductor materials emit different colors after electron injection.

A plasma display panel (PDP) uses the physical properties of a gas discharge. A cathode glow is produced when electrical current passes through small pockets of gas. Plasma displays can generally be divided into two major families, direct-current and alternating-current plasma displays. The color is typically orange on a black background.

A vacuum fluorescent display (VFD) produces light using rapidly moving electrons and phosphors in a vacuum under the control of a mesh grid. The VFDs are typically blue-green on a black background. Different colors are produced by combining different phosphors and using a wide range of filters.

An LCD is essentially a "light valve." The principle of the LCD is different from all of the other display devices discussed thus far. The LCD does not produce light but modifies it. When an electric field is applied, the alignment of the liquid crystal molecules in the display is altered. This alignment controls the polarization of light passing through the liquid crystal material by either blocking the light or allowing the light to pass though. LCDs can be classified into one of three types: reflective (ambient light), transmissive (backlit), or transflective (integration of transmissive and reflective). Both monochrome and color LCDs are available. Different addressing techniques have been developed, such as fast scanning, two-frequency addressing, hysteresis multiplex addressing, thermal addressing, and matrix addressing. The most popular addressing methods are passive matrix addressing and active matrix addressing. The first method uses a grid of wires (switches), whereas the second uses one to three transistors per pixel.

The low power consumption and compact nature of LCDs are major advantages. LCDs typically have power consumption of 2 to 3W (much less than a CRT). Lower driving voltage contributes to low power consumption in LCD products and to cost reduction of driving integrated circuits (ICs).

There are two types of color LCDs: Thin-film transistor (TFT) active matrix (AM) LCDs, and super-twisted nematic (STN) LCDs. At the same time, new breakthroughs are under way in development of glassless LCDs and low-temperature polysilicon TFT devices. A new method of liquid crystal patterning (which uses polarized ultraviolet light for patterning [17]) has been touted as very successful in increasing the viewing angles. Future LCD panels will see much improvements over the ones used in vehicles today. These improvements include but are not limited to the following [18]:

- Increasing the viewing angle by orientation control over the orientations of the liquid crystals: One such method involves using a multiple-domain technology to divide each pixel into two parts to double the viewing angle to ±40 degrees;

- Reducing the amount of reflection from the screen surface by using a black matrix and antiglare polarized panels;
- Reducing power consumption by developing better light-transmissive color filters, a wider aperture ratio, and highly efficient backlighting systems while improving brightness;
- Maintaining a lightweight, low profile by introducing highly reliable, high-density connections, a thinner backlight system, and lightweight materials;
- Development of a shock-proof structure by using new materials and production processes.

We now turn to flat panel displays. Flat panel displays are flat and light, and do not require a great deal of power. The ideal flat panel display is expected to be thin, have low volume, have an even surface, have high resolution, have high contrast, be readable in sunlight, display color, have low power consumption, and be solid state and lightweight. One example of this type of flat panel display is the field-emitter display (FED). Research and development on this technology have been very strong recently because LCDs, the technology with the largest market share, still produce an inferior image and have restricted viewing angles compared to CRTs.

The basic principle of the FED is similar to that of the CRT. In a CRT, there is one hot cathode that sprays electrons against the inside of the screen surface. In contrast, FEDs have millions of cold microtips that can be fired to turn pixels (picture elements) on and off (see Figure 7.1, where ITO stands for indium tin oxide). The FED has row and column conductors separated by an insulator layer. One conductive layer is the gate (extraction grid) and the other is the cathode (emitter). Each pixel is defined by an area where two conductive layers cross. Each pixel may have several (up to several hundred) microtips. Unlike active-matrix LCDs, which take one to three transistors to activate each pixel, the FED requires no transistors. A simple

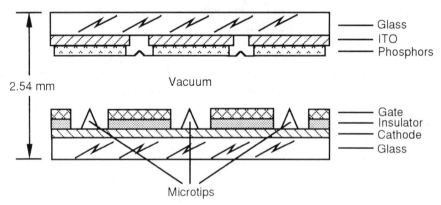

**Figure 7.1** Simplified cross-sectional view of FED layers.

wire grid powered by an electric voltage drives the firing of the microtips. Therefore, FEDs require less power than LCDs for an equivalent brightness. They look sharp from any angle, and are about 2.54 mm (one-tenth of an inch) thick, or about half the thickness of an LCD. FEDs have better or equivalent brightness, contrast, and resolution compared to both STN and AM LCDs. Furthermore, FEDs are 56% thinner, 65% lighter, have a 35% larger viewing area, and consume three to five times less power [19].

Interested readers will find a detailed discussion of the major visual display technologies (and flat panel displays) discussed in this section and the latest developments in [20–22] as well as the relevant display technology publications.

### 7.2.2 Touch Screens

Recent years have seen a steady increase in the amount of electronic equipment on vehicles, and the number of switches on the vehicle dashboard has made vehicles more complicated to operate. Reducing the number of input and output devices has therefore become an important design issue. One solution is to use a touch screen (panel), as a multifunction transparent switch to help simplify the input and output devices used in the vehicle.

As summarized in [12], using touch screens as control devices enables implementation of the following functional features:

• The ability to control a large number of functions from a small area;
• A potential reduction in the proliferation of conventional controls;
• Uniquely suited to provision pointing input in vehicles;
• Natural for menu selection;
• Best suited to occasional, low-resolution pointing.

To facilitate the usage of touch screens, general guidelines for the use of touch screens are provided in [5] and the relevant references therein. The touch-sensitive areas of the display should be indicated to the user. These areas should be at least 1.9 cm square and at least 0.32 cm apart. A touch screen should have sufficient luminance transmission so that the display is clearly readable in its intended environment and meets the 70 to 140 cd/m$^2$ luminance requirement, measured after any optical effects. Feedback should be provided to indicate touch screen actuation. A force-actuated touch screen should restrict the operating force to a value between 0.25 and 1.5 N. Only one touch command at a time should be accepted within approximately 100 ms, even though some technologies are capable of accepting more than one touch at the same time.

Touch screens are control devices. These screens are unique in that they can be overlaid on any flat CRT, LCD, ELD, or PDP panel. They are operated by the touch of a finger or stylus. The currently available technologies include resistive, capacitive, infrared, and acoustic wave. Since the resistive technology has the biggest market share, we discuss it in further detail in order to describe how it works and we also explain other implementation issues involved. Resistive touch screens are different from traditional devices because they are transparent, resistive, membrane switches. These screens are manufactured from thin, transparent, electrically conductive polyester films. Resistive touch screens are available in flat or curved configurations with either analog (continuously resistive) or matrix formats.

A matrix-format screen is divided into a grid of individual switches (or cells, see Figure 7.2). Touching an individual cell activates the corresponding switch. Matrix touch panels are relatively inexpensive, have relatively longer lives, and have lower cost electronic circuitry. However, the resolution is discrete rather than continuous.

The analog-format screen is not divided into a grid, but has a continuous resolution (Figure 7.3). The location on the screen that was touched is measured in both the X and Y directions. The resolution is determined by the decoding hardware and software. Analog touch panels have high resolution, provide quick response, and can be triggered by either nonconductive or conductive probes.

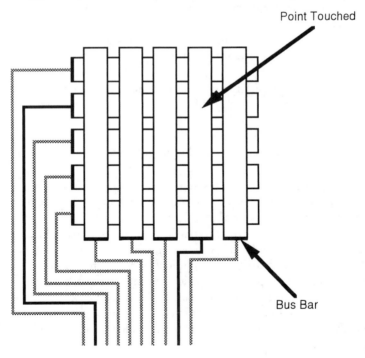

**Figure 7.2** Matrix-format touch panel.

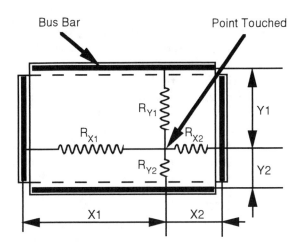

**Figure 7.3** Analog-format touch panel.

Even though touch panels looks very thin, they consist of many different layers. Figure 7.4 shows a simplified cross section through a touch panel on top of an LCD (in this diagram, ITO stands for indium tin oxide and PET stands for polyester). In our example, the two opposing transparent resistive (or conductive) layers consist of PET with sputtered ITO and glass with sputtered ITO separated by an insulating

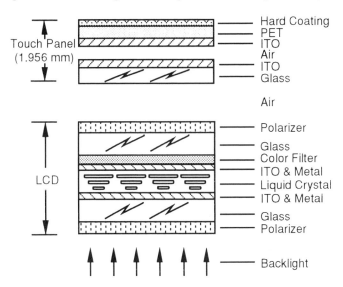

**Figure 7.4** Simplified cross-sectional view of the layers in a touch-screen panel and LCD.

layer (in Figure 7.4 this layer consists of air). The top substrate, in our case PET, should be flexible enough to allow actuation of the switch. The bottom substrate, in our case glass, can be either rigid or flexible. Note that for simplicity Figure 7.4 does not show the conductive bus bars.

Analog resistive touch panels operate by applying a voltage gradient across one conductive layer and measuring the voltage at the point of contact with the opposing conductive layer. Actuation with a finger or stylus on the touch panel brings the two opposing layers into electrical contact. Position information on the point of contact can be obtained by measuring the voltage levels. The conductive effects of these two resistive layers when activated are indicated by resistors in Figure 7.3. Even though the resistors on each axis appear on the same figure, they do not function simultaneously. For each decoding cycle, there are two scans, one along each axis. During each scan, only one layer is conductive and has a voltage gradient across it. The other layer acts as a high-impedance probe (sensor). Together these two layers act like a potentiometer with a sliding wiper. One layer is denoted as $X$ and the other is denoted as $Y$. The resistance primarily originates in the ITO sputtered surface and the silver-ink bus bars connected to the decoding circuitry. The ratio of the coordinates is proportional to the ratio of resistances in each direction:

$$\frac{R_{X1}}{R_{X2}} = \frac{X1}{X2}$$

$$\frac{R_{Y1}}{R_{Y2}} = \frac{Y1}{Y2}$$

The output voltage for the $X$ position may be calculated using

$$V_X \propto \frac{R_{X2}}{R_{X1} + R_{X2}}$$

where $V$ is the voltage and $R$ is the resistance. Similarly, the output voltage for the $Y$ position may be calculated from

$$V_Y \propto \frac{R_{Y2}}{R_{Y1} + R_{Y2}}$$

Since each touch panel may have slightly different characteristics, such as a different ratio of resistance, calibration is often necessary for accurate determination of the point touched. The touch panel can be calibrated first measuring the output voltages for known locations. Comparing these measurements to the expected values enables conversion factors to be determined. Because the actual output response and the expected response are proportional to one another, the conversion factors simply amount to a scale factor and an offset for each axis. Measurements of two points on each axis are required to determine the two conversion factors for each axis. Note that the calibration eliminates the major nonlinear effect due to the bus-bar resistance and can handle minor nonlinear perturbations such as that due to unevenly sputtered ITO. One calibration equation of this type is shown here:

$$P_{\text{out}} = \frac{P_A - P_B}{V_A - V_B} V_{\text{out}} - \frac{P_A - P_B}{V_A - V_B} V_{\text{offset}}$$

where $P$ is the position in either $X$ or $Y$, $V_{\text{offset}}$ is the voltage due to the driver and the bus-bar resistance. The terms $A$ and $B$ represent the two measured points, respectively. The calibration is also a function of the alignment between the touch panel and the display. If this alignment is not a constant, a dynamic calibration may be required. Moreover, when touch screens are being mounted on top of display screens, the distance between the touch panel and the surface of the display should be minimized in order to reduce parallax error.

At any boundary where there is a medium layer change, a certain amount of light is reflected. This causes glare on the screen. If the light is normal to the surface of the touch screen, the reflection can be calculated by

$$R = \left( \frac{n_1 - n_2}{n_1 + n_2} \right)^2$$

where $n_1$ is the index of refraction of one layer and $n_2$ is the index of refraction of the adjacent layer. From the preceding equation, we see that the most serious reflection occurs between layers having a large difference in the index of refraction and that more layers give rise to additional reflections. For the example shown in Figure 7.4, the indexes of refraction for PET, ITO, air, ITO, and glass are 1.67, 2.0, 1.0, 2.0, and 1.51, respectively. Knowing the thickness and index of refraction for each layer, the amount of light reflection and the transmittance of different touch-panel configurations can be calculated.

As the previous equation indicates, the extra resistive film added on top of the flat display panel for the resistive touch panel reduces the brightness or contrast and increases the amount of glare. One alternative to avoid this problem is to use an infrared touch technique, which has negligible effects on contrast. This technology is based on an infrared sensor array with scanned horizontal and vertical beams. These beams cross the surface of the display panel when they are turned on. Whenever both horizontal and vertical beams are blocked by a finger or stylus, a touch point is identified. The drawbacks of this technique are that it requires more computational power, costs more, has lower resolution (like the matrix format in Figure 7.2) and is more sensitive to the environment.

### 7.2.3   Additional Design Considerations

In addition to selecting a proper display device, one important factor that should be considered is that drivers often have difficulty reading detailed maps while driving. When a vehicle is in motion, it is preferable to display a limited amount of information such as a turn arrow, the shape of the next intersection, and the distance to the next intersection. When the vehicle is not in motion, a detailed map overlaid with the suggested route should provide a better overall picture of the surrounding area. For safety considerations, use of this map should be discouraged while the vehicle is in motion. In addition to variations on each display format [4,23], it is interesting to examine the two display formats, turn arrow versus route map, from the human factors perspective. One study has found that the route map makes drivers glance more often at the display screen than the turn arrow [9]. When assisted with voice guidance, the total number of glances is much smaller in both formats. In short, a turn arrow display with voice guidance should be used whenever possible.

As a design note, an interlocking mechanism, a dimmer mechanism, and a rotational mechanism (for displays that do not have a full viewing angle) should be included as control features for any visual-display-based interface. When the vehicle is moving, the interlocking mechanism should be able to automatically freeze most of the control functions for safety considerations. This is particularly important for touch-screen controlled displays. Interaction with the map display while driving should be discouraged. A dimmer mechanism can allow the driver to control the screen brightness. Otherwise, it might be very difficult for a driver to see outside clearly when switching from a very bright screen to a dark road at night. As an additional incentive to incorporate a dimmer, a U.S. National Highway Traffic Safety Administration regulation (NHTSA S5.3.5) requires this mechanism for any illumination source forward of a specified vertical transverse plane in the driver's compartment unless the light is more than barely discernible. The rotational mechanism for a limited-viewing-angle display is beneficial not only for the driver but also for any passengers who may wish to see the display screen. Some unpredictable light sources may make the screen very difficult to see from a particular angle. The ability to rotate the display can reduce these potential problems. Otherwise, a larger viewing angle display must be used.

The most important design considerations for an in-vehicle display are safety and ease of use. Eye glance frequency and duration of eye fixations on the display device are directly related to these two important design considerations. One study has found that the first-time user spent 20.4% of driving time looking at a poorly designed and complex route guidance device [24]. This reduced the glance time spent monitoring the road from 91% to 72.4%. Thus, this factor (which does not only apply to visual devices) must be kept in mind when designing the in-vehicle user interface. Because of the lack of adequate baseline data, it is still premature to establish guidelines or to provide a widely accepted set of numbers for glance frequency and duration. One study based on limited research has suggested the glance limits given in Table 7.3 [25] as a means of stimulating further discussions. Refer to this study and its references for further information on other human factor considerations, the methods used, and the data collected during field and laboratory research. More research is needed in this area.

Other design considerations for driver interfaces include when, where, and how the user should be informed of an upcoming turn (Section 6.2.2), what minimum text size should be used, how the graphical representation of the intersection should be designed, and so on. For instance, the most commonly used method of determining the character height is to specify that the visual angle subtended by a character (its height divided by the viewing distance) be greater than or equal to 0.007 radians. The turn-by-turn display should be simple and consistent while providing landmarks such as traffic lights and stop signs.

Before leaving the section devoted to visual-display-based interfaces, we want to briefly discuss a new software technology that has been developed recently [26] for displaying maps. It provides a simulated 3D view on the in-vehicle display monitor equivalent to that seen 1,000m from the rear of the car and at a height of 360m. The driver seems to "look down" at an angle of approximately 16 degrees, in the direction of travel (Figure 7.5).

In 2D display systems, it is often necessary for drivers to change the scale of their map displays manually, that is, to "zoom in" and "zoom out." For example, zooming out is commonly required in order to view the entire route in a large road network. And zooming in is often necessary when approaching an intersection. Because the 3D display can show a map with a foreground 500m wide and a background 7,000m wide [Figure 7.5(b)], it is easy to display the entire planned route on the screen without zooming features.

This 3D display was developed by converting a map database into a hypothetical image on which the vehicle position and planned route have been superimposed. It is designed to permit the section of the map closest to the current vehicle position to be shown in enlarged format. The scale of the display is continually updated as the trip progresses to show additional areas of the map. The 3D view is generated by turning the 2D map into a certain degree so the top edge of the original 2D map becomes a farground (back width, Figure 7.5) and the bottom edge of the original

Table 7.3
Glance Limits for Visual Display Devices

| Acceptability | Number of Glances | Glance Duration (sec) |
|---|---|---|
| Best expected | 2.0 | 0.9 |
| Desired/planned | 2.5 | 1.1 |
| Worst case | 4.0 | 1.2 |

(a)

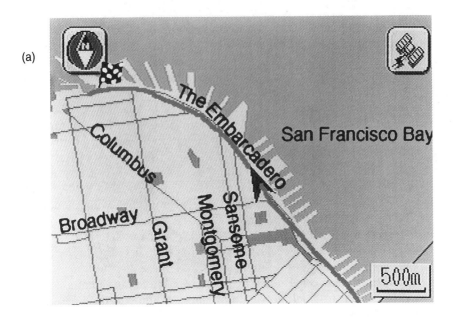

**Figure 7.5**  Comparison of (a) a 2D map (front and back widths: 500 m) and (b) a 3D map (front width: 500m; back width: 7,000m). (Courtesy of Nissan/Xanavi.)

2D map becomes a foreground (front width). Another technology that generates a true 3D display is discussed in [27]. Besides the 2D digital map database, it requires an additional database to derive altitude data.

## 7.3  VOICE-BASED INTERFACES

Speech is a natural interface that can be used to provide hands-free control of location and navigation systems. It is particularly important for the vehicle driver to operate the steering wheel rather than the navigation device during the trip. Because people typically speak three times faster than they can type, the access time for any computer device can be reduced, and access devices such as touch screens, push buttons, and keyboards can be eliminated using speech.

Table 7.4 lists two voice input and output methods, along with their evaluation scores and recommended usage [5]. The evaluation criteria used to grade these methods are identical to those used in Section 7.2. Upon comparing Tables 7.2 and 7.4, we see that voice input and output generally score higher than visual display technology for vehicle use.

(b)

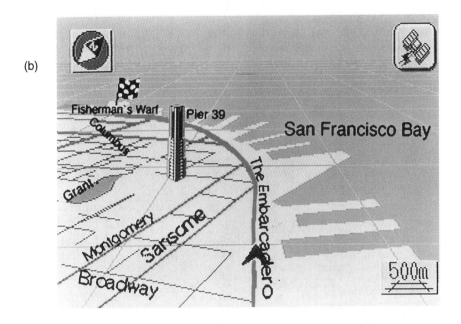

**Figure 7.5** (continued)

Table 7.4
Voice Input and Output Methods

| Voice | Score | Recommended Usage |
|-------|-------|-------------------|
| Speech | 8 | Where the user's eyes are busy or mobility is required |
| Auditory | 8 | When information is short, simple, and transitory, requiring an immediate or time-based response, visual display is restricted, etc. |

Voice processing methods can be classified as follows [28]:

- *Speech coding:* Obtaining compact digital representations of acoustic speech signals for efficient transmission or storage;
- *Speaker verification:* Verifying the claimed identity of a speaker for access to restricted areas or functions;
- *Speech recognition:* Recognizing an acoustic speech signal to identify the linguistic message intended;
- *Speech synthesis:* Generating an acoustic speech signal from ASCII text to convey a message from a machine to a human.

All four technologies can be used in vehicle location and navigation systems. A detailed review of these technologies can be found in the references listed in [28,29]. Speech coding can be used to convert human voice analog signals into coded digital signals [30]. When used for storage, speech encoding will maintain a desired level of voice quality while keeping the bit rate as low as possible. When used for transmission, search coding will conserve bandwidth or bit rate while maintaining adequate voice quality. Human voice signals can be recorded and stored in memory and then transmitted to a loudspeaker on demand for voice output used to guide the driver. Speaker verification can be used to add a security feature to the vehicle such as verifying the identity of persons attempting to access the vehicle. It works by determining whether an unknown speech sample was spoken by the individual whose identity was claimed. The process of speaker verification is similar to that of speech recognition. Speech recognition can be used as an input control device. Speech synthesis can be used as an output device. These two latter technologies are discussed in more detail in the next section.

### 7.3.1  Speech Recognition

Perfectly reliable speech recognition is difficult for a number of reasons. First, speech differs from person to person. Second, background noise often affects the voice signal. Third, people commonly make errors, slightly mispronounce words, insert extraneous words, or run words together when talking fast.

As summarized in [12], the following functions can be achieved when speech recognition is used for device controls:

- Hands-free control while driving;
- The availability of speaker-dependent and speaker-independent devices.

However, until confidence in speech recognition outcomes can be reliably demonstrated, it is better not to use it for critical inputs.

Speech recognition systems are often classified into three different types [28]. Systems of the first type are known as "isolated-word systems." In these systems, the user is expected to speak a single word or phrase from a specified vocabulary. The speech patterns used by these systems involve isolated words or phrases. Systems of the second type are known as "connected-word systems." In these systems, the user is expected to speak a fluent sequence of words and phrases, and the system matches each sequence to a string of words. The speech patterns used by these systems involve distinct words or phrases. Systems of the third type are known as "continuous-speech systems." In these systems the user is expected to speak fluently, and the system recognizes the speech based on word lexicon and an associated word grammar. The speech patterns of these systems are subword units from which words and phrases are created using a lexicon of word pronunciations.

Commercial speech recognition systems can also be classified as speaker dependent or speaker independent. Speech recognition systems can further be classified on the basis of discrete words versus continuous words, read text versus spontaneous speech, vocabulary size, etc. The low-end speaker-dependent technology is usually based on a template or acoustical representation of speech. User-trained "voice prints," or templates are stored in the system and then compared against the commands spoken by the user. Template-based speaker-dependent recognition is used for relatively small- to medium-sized vocabularies, occasionally up to a few hundred words. Another technique that uses speaker-dependent recognition systems involves matching subword units, such as phonemes, syllables, etc. This technique is often used for medium or large vocabulary sizes (in the thousands of words). Speaker-independent systems are supposed to work for any speaker, without training. The ideal goal is to develop a high-end continuous speaker-independent system with a large vocabulary, no required training, and high accuracy. In these systems, the subword units used in such systems are trained using a large number of speech samples that fairly represent the gender, accent, and dialect characteristics of the speaker. This large number of speech samples should provide a fair representation of the acoustic, phonetic, and linguistic variance within a recognition task or across several tasks. Despite recent advances, the best high-end system today is still impractical for embedded vehicle applications.

It is very common for vendors' products to use a statistical modeling technique based on hidden Markov models (HMMs) to match the features (parameters) extracted from human voice. HMMs are sophisticated statistical models. In the second-generation technology, HMMs are used to model subword units (phonemes), as opposed to words. HMMs can improve speech recognition performance in several ways. An HMM-based small-vocabulary system can tolerate wide variations in the analog speech signal and so can perform speaker-independent recognition, work reliably in high-noise environments, and recognize continuous speech. Alternatively, the added power of HMM technology can be used to distinguish a larger number of subtle differences between words, so that the recognizer can support a larger vocabulary.

Everything from small-vocabulary command and control using isolated words and small-vocabulary continuous speech through large-vocabulary dictation can be based on the same core HMM technology. In older products, spectral templates are used in place of HMM. Historically, the template-based method has not fared as well [29].

Many different algorithms and implementations exist for processing voice signals. The basic principle is to match the incoming transformed speech input against a stored representation. (This is often called the decoding process [29].) Acoustic models are generally used to capture the phonetic or word-level properties of speech while statistically based language models are often used to capture the syntactic and semantic regularities of language. The approximate diagram of Figure 7.6 provides a simple visual explanation of how a computer recognizes speech.

Current research has underlined the importance of robust and accurate front ends, detailed acoustic modeling, and statistical language modeling. Natural language processing techniques have also been studied for speech interface and interpretation of spontaneous speech. Combinations of neural networks for signal classification and HMMs for speech modeling have been demonstrated successfully.

Recent products for recognition of continuous speech typically have maximum vocabularies of around 2,000 words, with a *perplexity* of 30. Perplexity is a measure of sentence complexity corresponding to the average number of words in the vocabulary that could follow any given word under the recognizer's syntax rules. A typical *Wall Street Journal* article has perplexities of around 200 to 300. Thus, there is very clearly a large gap between user expectations and reality for current speech recognition products.

Some researchers believe that the most important challenge is the integration of speech recognition and natural language processing technology to achieve speech understanding. They have attempted to build a probabilistic parser that operates top-down, using a stack-based control strategy that favors the most likely analyses. It

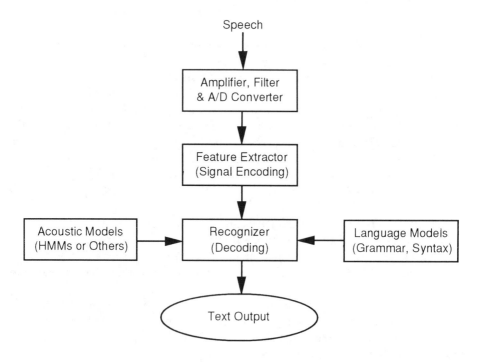

**Figure 7.6** Simple block diagram for speech recognition.

is designed so that its grammar rules and associated probabilities can be automatically trained from a set of parable sentences. The framework is organized such that the prediction probabilities for next-word candidates are explicit. This approach is different from the way probabilities are handled in other probabilistic natural language frameworks, where probabilities are typically associated with rule predictions.

For practical systems, large vocabulary systems must be made robust against interfering noise. Two types of methods are being studied. First, the corrupted speech waveform may be preprocessed so that the resulting parameters are very similar to those of clean speech. Techniques in this category include spectral subtraction, spectral mapping, and inherently robust parameterization. In the first category, some schemes have also attempted to estimate the clean speech signal using information about the speech. Their representatives are HMMs and minimum mean square error estimators. The second group of methods attempts to adjust the pattern-matching stage to compensate for the interfering noise. Methods using this approach include noise masking, state-based filtering, cepstral mean compensation, HMM decomposition, and parallel model combination.

Many other techniques (such as fuzzy logic and artificial neural networks) are being investigated by researchers around the world. Only time will tell if these techniques can be successful in the real-world environment.

Even when the host CPU has enough horsepower to run the entire speech recognition algorithm, a digital signal processor (DSP) has distinct cost advantages, especially in continuous-speech systems. Low-cost DSPs replace the expensive analog hardware formerly required to perform high-quality speech filtering and digitization. High-end DSPs can also replace the conventional CPUs.

The trend of DSP development is that the performance of DSP chips doubles about every 3 years, while the costs decrease at approximately the same rate. All commercial speech recognition algorithms can run on general-purpose platforms, but sometimes it pays to optimize the hardware for the task at hand, particularly in cost-sensitive, high-volume consumer applications. Therefore, there has been a trend to port speech recognition technology to specialized DSPs for products such as personal digital assistance (PDA) and other hand-held or mobile platforms.

Because there are so many speech recognition products available, one might consider integrating these products into various vehicle location and navigation systems. Generally, most of the hardware cost is in the memory and in the audio front end. The cost of the software varies over a very wide range. High-volume applications can bring per-unit licensing costs down to a very competitive price (less than $100, perhaps even less than $10). The best way to keep the cost down is to minimize the need for modifications to the vendors' original designs.

A few general guidelines for design of spoken input are available (see [4] and the references contained therein). The input should be easy to learn and remember, and should be controlled by brief commands. It should present explicit feedback describing whether the system is ready to accept speech or not. Acknowledgment should be given for each user command in understandable phrases; the feedback should be context sensitive, and distinguish between data and commands. Driver speech inputs should be acoustically distinguishable. Command names should be chosen so as to be semantically discriminable. The speech recognition devices should be able to accept a variety of pronunciations.

Development of advanced speech recognition technology requires expertise in signal processing, system theory, statistics, pattern recognition, and computer science, built on a solid understanding of speech science and linguistics. This is especially challenging for vehicle-based applications. As a rule of thumb, a speech recognition product is ready to be used in an application once the word error rate has been reduced below about 5% in its intended environment. Unfortunately, this is still not reliable enough for any critical control input to a vehicle during an emergency situation.

### 7.3.2 Speech Synthesis

Speech synthesis is a computer-aided technology output understandable human language. As summarized in [12], the following functions can be achieved using speech output:

- Wide range of possible output information;
- Low visual demand;
- Low training or skill demand;
- Best for short, simple messages.

However, the speech output is poor for visual or spatial information. Complex information cannot be extracted as easily as in a visual display. If speech synthesis is overused, the driver could be distracted.

Speech synthesis systems can convert text messages to speech based on a linguistic analysis of the text and an acoustic analysis of the speech sounds within the context of the desired sentences (Figure 7.7). These sounds can be phonemes, syllables, words, etc. Intonations, word pronunciations, exceptions, and rules for generating contextual changes in the sounds are used in the linguistic and acoustic analysis. Finally, the speech output can be obtained by generating appropriate word durations as well as global and local pitch contours for the prominence of words and ideas within the text. Hopefully, the output will be intelligible and reasonably natural.

Two synthesis techniques (as shown in Figure 7.7) are commonly used [29]. Formant synthesis uses a mathematical model of the waveform generated by the human speech production apparatus. Concatenation synthesis splices prerecorded segments of speech. The first technique uses sound symbols to define a sequence of

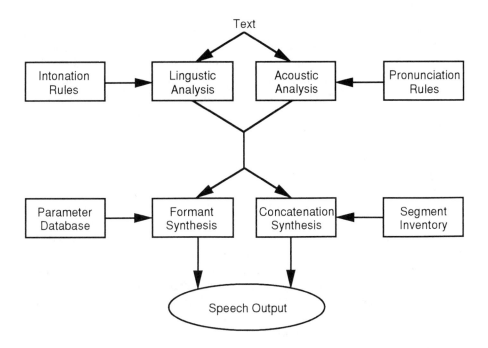

**Figure 7.7** Simple block diagram for speech synthesis.

acoustic targets. It interpolates the acoustic signal between these targets to mimic the dynamics of the human voice. The second technique obtains the same effect by using a sequence of prerecorded elements corresponding to the sound symbols in context. It smooths the junctures between these elements. Because concatenation synthesis is based on real human voices, it can produce a more natural voice, although more storage space is required (unless CPU-intensive speech coding is performed).

Progress in synthesizers appears to have reached an asymptote because successive small improvements are now difficult to achieve [29]. Although speech synthesis is currently a stable technology with error rates as low as 3.25% for individual word perception on a standardized test, the available synthesizers still do not sound natural. Because formant synthesis seems to have reached a plateau, any dramatic improvements in the future are expected to depend on concatenative synthesizers.

A few general guidelines concerning auditory output are available (see [4] and references contained therein). Nonspeech auditory sounds should be used only for providing alerts or warnings. Other messages should be spoken. Synthesized speech is only recommended for situations that require substantial flexibility in generating the spoken messages. When the messages are relatively limited and known ahead of time, a recorded human voice is preferred. Instructions should always be limited to four chunks (three or less in most cases). For a maneuver instruction, it is better to provide "early," "prepare," and "approach" messages in sequence (see Section 6.2). Each decision should be accompanied by a separate instruction when time permits. Sometimes, an auditory output may be required to remind the driver that the system is still operational. When the street name changes or a new city is entered, proper instructions should be announced. A function to activate and deactivate auditory output and to adjust the output volume should be provided.

Both synthesized and prerecorded (digitized or coded) voice output have been shown to be very useful in vehicle guidance. As mentioned earlier, studies of voice guidance systems have found reductions in the duration and frequency of glancing at the display relative to nonvoice guidance systems [9,31]. This can certainly help improve safety.

As we have learned, speech synthesis has become a stable technology. There are many examples describing the use of this technology [32,33]. Speech recognition technology has even recently become available in vehicle navigation products. One example of this is a speaker-independent voice-activated navigation system called AudioNav [34]. It recognizes isolated letters, numbers, and three dozen words and has a synthesized voice output. This system uses HMM technology and claims to be sufficiently robust for use in this important, but difficult, application environment—the automobile. Road noise and the impracticality of using a headset microphone while driving make automotive speech recognition a particular challenge. This fact is reflected in AudioNav. One main disadvantage of AudioNav is the inconvenience of voice input. It requires the user to spell out the destination and requires the user to know his or her present vehicle location.

Suppose that future speech processing technologies are natural and robust enough for vehicle applications. Will drivers prefer voice devices over visual displays? If both visual and voice devices are integrated in a vehicle location and navigation system, what is the optimum combination of visual and voice interfaces to maximize the safety and performance for both the system and the human? Will a route guidance system ever be better than a human travel guide who is familiar with the area [35]?

## References

[1] M. S. Sanders and E. J. McCormick, *Human Factors in Engineering and Design*, 7th ed., New York: McGraw-Hill, 1993.

[2] D. A. Norman, *Psychology of Everyday Things*, New York: Doubleday, 1988.

[3] B. Peacock and W. Karwowski, *Automotive Ergonomics*, Washington, DC: Taylor & Francis, 1993.

[4] P. Green, W. Levison, G. Paelke, and C. Serafin, *Preliminary Human Factors Design Guidelines for Driver Information Systems*, Ann Arbor, MI: The University of Michigan, Transportation Research Institute, Technical Report No. UMTRI-93-21. Also published in McLean, VA: U. S. Dept. of Transportation, Federal Highway Administration, Report No. FHWA-RD-94-087, Dec. 1995.

[5] L. Levitan, M. Burrus, W. L. Dewing, W. F. Reinhart, P. Vora, and R. E. Llaneras, *Preliminary Human Factors Design Guidelines for Automated Highway System Designers*, McLean, VA: U.S.

Dept. of Transportation, Federal Highway Administration, Report No. FHWA-RD-94-116, Dec. 1995.

[6] Department of Defense, *Handbook for Human Engineering Design Guidelines*, MIL-HDBK-759C, Philadelphia, PA: Department of Defense Single Stock Point—Customer Service, July 1995.

[7] K. H. E. Kroemer, H. B. Kroemer, and K. E. Kroemer-Elbert, *Ergonomics: How to Design for Ease and Efficiency*, Englewood Cliffs, NJ: Prentice Hall, 1994.

[8] D. J. Mayhew, *Principles and Guidelines in Software User Interface Design*, Englewood Cliffs, NJ: Prentice Hall, 1992.

[9] M. C. Hulse, T. A. Dingus, D. V. McGehee, and R. N. Fleischman, "The Effects of Area Familiarity and Navigation Method on ATIS Use," *Proc. Human Factors and Ergonomics Society 39th Annual Meeting*, Oct. 1995, pp. 1068–1071.

[10] W. B. Verwy and N. A. Kaptein, "Automatic Man-Machine Interface Adaptation to Driver Workload," *Proc. First World Congress on Applications of Transport Telematics and Intelligent Vehicle-Highway Systems*, Nov./Dec. 1994, pp. 1701–1708.

[11] J. Walker, E. Alicandri, C. Sedney, and K. Roberts, "In-Vehicle Navigation Devices: Effects on the Safety of Driver Performance," *Proc. IEEE Vehicle Navigation and Information Systems Conference (VNIS '91)*, Oct. 1991, pp. 499–525.

[12] B. Leiser and K. Carr (Eds.), *Analysis of Input and Output Devices for In-Car Use*, Traffic Research Center, University of Groningen, The Netherlands, Nov. 1991.

[13] A. Nojima, H. Morita, H. Kondo, and N. Komoda, "Development of Toyota ATIS In-Vehicle Equipment in VICS Demonstration Test," *Proc. IVHS America 1994 Annual Meeting*, Apr. 1994, pp. 664–670.

[14] A. Marcus, "Motorola Smart Car," in *Macromedia Director Design Guide*, C. Clarke and L. Swearingen, Eds., Indianapolis, IN: Hayden Books, 1994, pp. 66–76.

[15] M. F. Coleman, B. A. Loring, and M. E. Wiklund, "User Performance on Typing Tasks Involving Reduced-Size, Touch Screen Keyboards," *Proc. IEEE Vehicle Navigation and Information Systems Conference (VNIS '91)*, Oct. 1991, pp. 543–549.

[16] S. L. Fowler and V. R. Stanwick, *The GUI Style Guide*, Boston: AP Professional, 1995.

[17] M. Schadt, H. Seiberle, and A. Schuster, "Optical Patterning of Multi-Domain Liquid-Crystal Displays with Wide Viewing Angles," *Nature*, Vol. 381, May 16, 1996, pp. 212–215.

[18] "Sharp Realizes New Applications with Technology Breakthroughs," *Display Devices*, No. 11, Spring 1995, pp. 12–13.

[19] C. Ajluni, "FED Technology Takes Display Industry by Storm," *Electron. Des.*, Oct. 25, 1994, pp. 56–66.

[20] S. Matsumoto (Ed.), *Electronic Display Devices*, translated by F. R. D. Apps, New York: John Wiley & Sons, 1990.

[21] R. A. Perez, *Electronic Display Devices*, Blue Ridge Summit, PA: TAB Books, 1988.

[22] L. E. Tannas, Jr. (Ed.), *Flat-Panel Displays and CRTs*, New York: Van Nostrand Reinhold, 1985.

[23] J. M. C. Schraagen, "Information Presentation in In-Car Navigation Systems," in *Driving Future Vehicles*, A. M. Parkes and S. Franzen (Eds.), Washington, DC: Taylor & Francis, 1993, pp. 171–185.

[24] T. Ross and G. Burnett, "The Right Road to Take," *ITS: intelligent transport systems*, June 1996, pp. 54–55.

[25] P. Green, *Suggested Procedures and Acceptance Limits for Assessing the Safety and Ease of Use of Driver Information Systems*, Ann Arbor, MI: The University of Michigan, Transportation Research Institute, Technical Report No. UMTRI-93-13. Also published in McLean, VA: U. S. Dept. of Transportation, Federal Highway Administration, Report No. FHWA-RD-94-089, Dec. 1995.

[26] M. Watanabe, O. Nakayama, and N. Kishi, "Development and Evaluation of a Car Navigation System Providing a Bird's-Eye View Map Display, *SAE Paper No. 961007*, Society of Automotive Engineers, 1996.

[27] M. Oikawa and M. Kato, "A Three Dimensional Navigation System '3D Navi'," *Abs. Third World Congress on Intelligent Transport Systems,* Paper #3129 (full paper in CD-ROM), Oct. 1996, p. 205.

[28] L. R. Rabiner, "The Impact of Voice Processing on Modern Telecommunications," *Speech Comm.,* Vol. 17, Nos. 3-4, Nov. 1995, pp. 217–226.

[29] A. I. Rudnicky, A. G. Hauptmann, and K.-F. Lee, "Survey of Current Speech Technology," *Comm.* ACM, Vol. 37, No. 3, Mar. 1994, pp. 52–57.

[30] A. S. Spanias, "Speech Coding: A Tutorial Review," *Proc. IEEE,* Vol. 82, No. 10, Oct. 1994, pp. 1541–1582.

[31] H. Kishi and S. Sugiura, "Human Factors Considerations for Voice Route Guidance," *SAE Paper No. 930553,* Society of Automotive Engineers, 1993.

[32] J. R. Davis and C. M. Schmandt, "The Back Seat Driver: Real Time Spoken Driving Instructions," *Proc. IEEE Vehicle Navigation and Information Systems Conference (VNIS '89),* Sep. 1989, pp. 146–150.

[33] M. K. Krage, "The TravTeck Driver Information System," *Proc. IEEE Vehicle Navigation and Information Systems Conference (VNIS '91),* Oct. 1991, pp. 739–748.

[34] J. Smolders, R. Diller, and D. Van Compernolle, "Noise Robust Speech Recognition Makes In-Car Navigation Safe and Affordable," *Proc. Second World Congress on Intelligent Transport Systems,* Nov. 1995, pp. 601–604.

[35] W. Fastenmeier, R. Haller, and G. Lerner, "Preliminary Safety Evaluation of Route Guidance Comparing Different MMI Concepts," *Proc. First World Congress on ITS,* Nov. 1994, pp. 1750–1757.

# CHAPTER 8
▼▼▼

# WIRELESS COMMUNICATIONS MODULE

## 8.1 INTRODUCTION

Reliable communication is a vital component for further improvement of performance and increased functionality in vehicle location and navigation systems. It provides a very valuable opportunity to present relevant information (and invaluable safety features) to the vehicle and its occupants as well as to obtain data for transportation management systems. Many quality services can be provided to drivers using communications technology. For example, the drivers and in-vehicle navigation systems can receive updated traffic information to help them maneuver through traffic, the traffic control centers can obtain current traffic reports from vehicles for traffic management and travel time predictions, and traffic control centers could even locate and help the vehicle navigate in the road network.

Communications technologies are poised for major expansion in two key areas: wireless data communications and high-speed wireline (wire-based) communications supported by the Internet and ancillary networks. Ultimately it will be the marriage of these two key technologies that will bring about a revolution resulting in a new information society.

Wireless data applications play a critical role in making the vision of mobile computing a reality. Today's competitive and fast-paced business climate demands tools that allow people to communicate at their own convenience and discretion. Wireless applications continue to be refined and will become commonplace in the near future. Data systems could develop into one of the fastest growing sectors of the wireless industry.

High-speed wireline infrastructures such as the Internet are a key element to providing a backbone for data delivery. This infrastructure has approximately doubled in size each year since 1988. A key factor to this dramatic growth has been the Internet's widespread connectivity and universal access, which often makes it quicker and easier for people to route their traffic through a local access point than to talk directly to each other. A national backbone with countless local access points gives the mobile community easy access to vast amounts of information and services not available on most private networks. This access has propelled the capabilities of wireless computing into the global arena.

The emphasis in this chapter falls more on wireless communications, and most of the technologies to be discussed are based on radio-frequency (RF) communications. Wireline communications are discussed spontaneously in connection with the wireless technologies when appropriate. To concentrate on our main topic of describing the existing communications technologies, we will not discuss how to use the radio signal for positioning here. However, we should keep in mind that many of the technologies discussed can be used for positioning of vehicles or mobile devices, which is addressed in Section 9.2. A large number of diverse technologies are available and space is limited, so it would be difficult to address all of these technologies in a single chapter. We therefore discuss only the major technologies currently used for vehicle location and navigation applications and direct readers to the references listed at the end of this chapter for further reading on other technologies. Section 8.2 discusses the basic attributes of communication subsystems. Section 8.3 provides an overview of each technology to quickly bring the reader up to date, but also contains important details, many useful summary tables, and comprehensive references to encourage further research.

## 8.2 COMMUNICATIONS SUBSYSTEM ATTRIBUTES

Vehicle communications require a seamless, wireless infrastructure for voice and data that can reliably and efficiently deliver real-time traffic and other information between a fixed infrastructure (base stations or repeaters) and mobile vehicles (mobile radios) or mobile devices (portable radios) or between the mobile vehicles themselves. Depending on whether the link is simplex (information is transmitted in one direction or both directions by alternating between transmitting and receiving on the same frequency), half-duplex (information is transmitted in both directions by alternating between transmitting and receiving but on different frequencies), duplex (information is transmitted in both directions simultaneously using different frequencies), a matching transmitter and/or receiver must be used both in the mobile vehicles and in the infrastructure. In this book, when referring to two-way communication, we will mean either duplex or half-duplex depending on the context. Mobile communications can be both one-to-many and one-to-one. Broadcasting is a one-way, one-to-many communication, in which a single message can be received by all mobile receivers within range. For one-to-one communication, a communications channels (frequency) is reserved solely for use by two parties for the duration of the communication. A dispatch system provides two-way communications channels for a designated group (or fleet) of users. As is apparent, different applications have diverse communication requirements. Furthermore, certain key attributes are required for successful deployment. All of these requirements and attributes apply equally to the intelligent transportation system (ITS), which is an emerging mobile computing technology. For ideal widespread ITS deployment, these attributes should include excellent coverage, high capacity, low cost, full connectivity, excellent security, and public access. We now briefly discuss the key attributes that can be used to assist in the selection of proper technologies for vehicle applications.

### 8.2.1 Coverage

Coverage refers to the geographic area where wireless voice and data transmissions can be sent and received reliably. Specific site or station coverage may depend on a number of factors at both the fixed and mobile ends. Transmitter power, antenna characteristics, channel characteristics, receiver quality, interference, communications protocols, and geographic topography are generally the factors that affect site coverage the most.

   Boosting transmitter power is one method that can be used to increase coverage, but the benefits are limited since most base stations/repeaters typically have a transmitter power of 75 to 100W. On the mobile or portable end, it is desirable to obtain the required coverage with a minimum amount of transmitter power. This is because the transmitter power of portables is usually limited to 3 to 5W. To overcome the limited transmitter power of portable radios, tower-top receiver amplifiers are often added to the base stations/repeaters for better signal reception. In any case, doubling the input power results in only a 40% increase in field strength. Proper selection of antenna characteristics can be more effective in increasing coverage. At the base station end, increasing the antenna height is usually a very good solution. However, antenna towers are expensive and their heights are often restricted by local regulations.

   Gain and height are two of the most important characteristics of an antenna. Antenna gain is a measure of how well the antenna concentrates its radiated power in a given direction. It provides a measure of the antenna's efficiency. The gain of an antenna is measured as the ratio between the highest radiation intensity (in any direction) from the antenna and the radiation intensity from a reference antenna (such as a dipole antenna) or a theoretical ominidirectional antenna driven by the same total power, that is, an isotropic source that radiates energy in all directions. Antenna gain is measured in decibels (dBs) relative to an isotropic antenna, that is, dBi, and may vary widely depending on the application and design of the antenna. Gain numbers can range from 0 to 50 dB (for example, 0 to 17 dB for antennas that are omnidirectional in the horizontal plane).

The second main characteristic of an antenna is its height. This explains why most transmitters are located at the tops of tall buildings or hills to get maximum coverage. If no buildings are available, a tower is sometimes constructed along with a protected enclosure to house the base station equipment. Optimum height is determined by the required coverage, range, antenna characteristics, local regulations, the trade-off between extra height and greater signal attenuation (loss), and cost. To keep signal losses in the transmission lines to a minimum, the base station equipment is generally kept as close as possible to the antenna.

Table 8.1 summarizes the properties of antennas in each frequency band, assuming that the noise density is higher in urban areas and interference is present from another antenna in the same area. This table gives us a broad picture of the relationship between frequency and coverage. Note that even though a high-gain antenna design can be more easily constructed at higher frequencies, too much directivity may not be acceptable for the application, and path loss is inversely proportional to the wavelength squared.

As Table 8.1 implies, channel characteristics such as path loss, fading, and penetration loss can seriously impact coverage. Path loss is the attenuation of the signal as it propagates from the transmitting antenna to the receiving antenna. In practice, the path loss is more severe than the inverse square law would predict from free-space calculations. The path loss for mobile systems generally falls off as distance is cubed (or even faster). However, it is important to note that these figures must often be empirically derived.

**Table 8.1**
Frequency Band Characteristics

| Characteristics | 25–50 MHz | 148–174 MHz | 450–512 MHz | 806–960 MHz |
|---|---|---|---|---|
| Antenna size | Long | Short | Shorter | Shortest |
| Antenna gain | Low gain | Low–high gain | Low–high gain | Low–high gain |
| Urban range | Poor | Good | Excellent | Excellent |
| Suburban range | Good | Excellent | Good | Fair |
| Rural range | Excellent | Good | Fair | Poor |
| Urban building "fill-in" | Poor | Fair | Good | Excellent |
| Foliage path loss | Low | Low | Medium | High |
| Interference | Severe | Medium | Minimal | Almost none |

As the signal travels along its transmission path, it passes through air and other environmental material. Oxygen and water vapor in the air may absorb energy, and the resulting attenuation depends mainly on the frequency, which means that the attenuation due to rain increases with frequency and amount of precipitation. Frequencies below 1 GHz are essentially unaffected by atmospheric moisture, while frequencies above 10 GHz are severely affected (or even become unusable) for long transmission paths. Finally as Table 8.1 indicates, the 806- to 960-MHz band has high foliage attenuation, but good urban fill-in coverage (between buildings). The 148- to 174-MHz band has low foliage attenuation, but poor urban fill-in coverage.

Fading is a phenomenon that leads to variations in signal level. A distinction is made between short-term and long-term fading. For the short-term fading, signals vary over short distances due to multipath propagation. This type of fading generally becomes more severe as the frequency increases. Long-term fading is a random process caused by shadowing. For radio frequencies of 450 to 960 MHz or higher, shadowing becomes the most important environmental attenuation effect where buildings or other ground terrain may create radio shadows that can result in received signal strength differences as great as 20 dB. These fading effects are most severe in dense metropolitan areas. On the other hand, high frequencies bounce off buildings, providing good fill-in coverage (as mentioned above), although the ability of the signals to penetrate through buildings is a function of frequency band and signal strength.

Receiver sensitivity is an important measure of receiver quality, and has been defined in various ways depending on the type of signal modulation. Modulation is a technique for transmitting information by varying the amplitude, frequency, or phase of a carrier wave or signal. Sensitivity is a measure of the smallest signal that can be successfully received. A receiver with high sensitivity is desirable, since a lower field strength is required for the receiver to respond correctly at the fringe of the coverage area. For the analog signals used by FM (frequency modulation) receivers, the oldest standard for sensitivity measurement standard uses the 20-dB quieting method. In this method, the sensitivity is measured by determining the unmodulated input signal level that results in the noise output of the receiver being reduced 20 dB below the noise output level with no input signal. Another widely accepted measure of the sensitivity uses the 12-dB SINAD method [SINAD refers to the ratio of (signal + noise + distortion) to (noise + distortion) expressed in decibels]. Digital communication systems specify the sensitivity in terms of the signal level required to attain a bit error rate (BER) not exceeding 1%, or in many situations, the maximum BER allowed for acceptable system operation. Signal strength generally is given in microvolts; however, in some situations it will be specified in decibels relative to one milliwatt (dBm).

Features such as error detection and correction, data interleaving and coding, and the retry control employed by the communications protocol may directly affect coverage. A trade-off is usually necessary between throughput and coverage in this case. Error detection and correction work by adding information to the message (sometimes referred to as *redundancy* when multiple copies of certain data are transmitted). This method can be used to detect, and perhaps correct, errors in transmission. The more redundancy added to the message, the more reliably the receiver can detect and correct errors, but with a corresponding reduction in the effective message rate. In practice, the number of additional check digits required to achieve error correction is much larger than that needed for just error detection. Convolutional encoding may be used in the forward error correction process to produce an output bit stream representing the original message such that the number of bits between adjacent bits from the original message is larger than the number of bits affected by a single burst error. For example, if an error burst of up to seven bits in length does occur, it will affect only a single bit in each codeword rather than a string of bits in one or two codewords. If an error is detected but cannot be corrected, then an automatic repeat request (ARQ) can be transmitted back to the sending unit to retry the transmission. Another alternative is the selective ARQ (SARQ), which requests retransmission of only the corrupted piece of information rather than the entire digital transmission. Thus, error detection and correction can result in a lower end-to-end BER for the message as a whole.

Geometry and topography must be taken into account when determining site coverage because hills, buildings, and other geographic features may absorb, refract, or reflect signals. Many coverage prediction models contain detailed geographic data, including the locations of buildings and terrain features to be used in coverage calculations.

The overall network coverage depends on the concentration and distribution of network equipment such as base stations. For example, cellular services typically position their base stations to emphasize wide geographical coverage. At the other extreme, infrared and microwave equipment generally requires line of sight between transmitter and receiver. Some packet data network operators concentrate their base stations in urban areas to emphasize in-building coverage. Such mobile technologies typically only serve metropolitan areas where system operators have the best opportunity for recovering their investment. On the other hand, the requirement for wide-area coverage is important if seamless nationwide operation of ITS is to be provided.

### 8.2.2 Capacity

Capacity varies in terms of message size and effective throughput. For example, paging services generally have limited message size. The effective throughput depends on transmission speed and number of concurrent transmissions/connections. Although capacity is directly related to transmission speed, the transmission speed figures cited by wireless providers can be misleading. Protocol overhead associated with error correction and reliable message delivery can account for throughput rates that are as much as 50% lower than the maximum over-the-air transmission rate. Also, some mobile technologies employ store-and-forward delivery to compensate for their limited radio bandwidth or service capacity. With such networks, messages can be delayed anywhere from a few seconds to hours but may have substantially lower operating costs. In general, communications technologies with higher capacity and throughput will have higher operating costs.

### 8.2.3 Cost

Equipment and operating costs vary widely according to device capabilities, transmission rates, device size, and power requirements. Paging technology is generally the least expensive (and least capable) technology to implement, whereas satellite and microwave technologies are the most expensive to deploy. To provide universal access to ITS as a national network, subscriber costs must be kept to a minimum.

### 8.2.4 Connectivity

Connectivity may be supported by simplex, duplex, and/or half-duplex communications to provide broadcast, dispatch (group), and individual communications within the network. Simplex connectivity can be used for simple data collection purposes where sensor data is delivered to a processing center. Two-way communications are usually required for interactive sessions. A mobile or portable radio user may converse with the communications center, an individual radio user, a group of vehicles, or the whole network to communicate and exchange information. The situation is similar for the operator in a communications center.

## 8.3 EXISTING COMMUNICATIONS TECHNOLOGIES

Vehicle communications networks need to support many different applications within a wide set of user services including traffic management, emergency management, public transportation operations, commercial vehicle operations, electronic payment, and advanced vehicle control. Each application has specific needs that may best be satisfied by a particular communications technology. Because no single communications technology can satisfy all of these requirements, a hybrid implementation of technologies is likely to be required for a regional or statewide communications network. Table 8.2 provides a high-level view of existing wireless technologies (after [1]). This matrix is intended as an initial guide for use at the beginning of the selection process. The user should perform a more thorough evaluation of the planned application and available technology prior to actual implementation of the system. Many good references are available to aid the user with this analysis (see [1,2] and the references contained therein).

**Table 8.2**
Communications Technology Matrix

| Technology | Reliability | Coverage | Avg. Data Transfer Speed | Equipment and Airtime Costs | Security | Simplex or Duplex | Real Time or Store and Forward | Access |
|---|---|---|---|---|---|---|---|---|
| Paging | E | Metro, some rural | 2.4–3.6 Kbps | E | E | Simplex/duplex | Store and Forward/Real time | Public |
| Cellular | F-E | Metro, some rural | 9.6–19.2 Kbps | F-E | P-E | Duplex | Real time | Public |
| PCS | E | Metro | ≥8 Kbps | E | E | Duplex | Real time | Public |
| Private land mobile radio systems | VG-E | Metro, rural | 1.2–64 Kbps | F-E | P-E | Simplex/duplex | Real time | Private/public |
| Radio data networks | VG-E | Metro | 2.4–256 Kbps | F-E | E | Duplex | Real time | Public |
| Broadcast subcarriers | E | Metro | 1.2–19 Kbps | E | E | Simplex | Real time | Private/public |
| Short-range beacons | E | 1–100m | 64–1024 Kbps | F | VG | Simplex/duplex | Real time | Private |
| Satellites | E | Worldwide | 2.4–21.33 Kbps | P-E | E | Duplex | Real time | Private |
| Cordless telephony | E | None in the U.S. | ≥32 Kbps | VG | E | Duplex | Real time | Public |
| Radio LAN networks | VG-E | Indoors/outdoors; 40–11,263m | 64–5,700 Kbps | VG | E | CSMA/CD | Near real time | Private |
| Infrared LAN networks | E | Indoors/Outdoors 9–6,436m | 1,000–10,000 Kbps | F-E | E | CSMA/CD | Near real time | Private |
| Meteor burst | E | Worldwide | 2–32 Kbps | E | E | Duplex | Store and forward | Private |
| Microwave relays | VG-E | Metro, rural; 40,000-m hops | 8448–250 × 1544 Kbps | VG-E | VG-E | Duplex | Real time | Private |

Note: P = poor; F = fair; VG = very good; E = excellent; Metro = metropolitan areas; CSMA/CD = carrier sense multiple access with collision detection.

Both analog and digital systems are available for mobile radios (the digital systems are essentially related to voice digitization and to digital transmission). The major advantages of the digital systems over the analog systems are increased spectrum efficiency, more consistent audio quality throughout more of the coverage area, inherent privacy from analog scanners, and greater data rate capability.

We now take a brief look at some of the transmission and switching methods used for transmission of information. Three major channel access methods exist: FDMA (frequency-division multiple access), TDMA (time-division multiple access), and CDMA (code-division multiple access) [3]. FDMA and TDMA can be implemented on either narrowband channels (e.g., 12.5 kHz) or wideband channels (e.g., 25/30 kHz), whereas CDMA is restricted to the wideband architecture. Two principal switching techniques are available: circuit switching and packet switching. Circuit switching requires a complete end-to-end connection over both the wireless and wired segments before any voice or data can be sent. Packet switching divides the information into packets and transfers packets of data between network nodes over different connections until the packets reach their final destination and are reconstructed for the user. The main difference between these two methods is that circuit switching reserves the required bandwidth in advance, whereas packet switching acquires and releases it on an as-needed basis. Two other switching techniques are store-and-forward (in which the information is first stored and then forwarded, one hop at a time) and hybrid (variants and mixed forms of circuit switching and packet switching) transmission.

In general, mobile services can be classified into conventional (nontrunked) mobile systems and trunked mobile systems. Each conventional mobile system has one channel available for each specific group of users. In other words, users are grouped so each group of users is assigned to a different channel. This system is less efficient than trunking because users can only communicate if their channel is free, regardless of whether the other channels are free. Since each group of users shares a common radio frequency, the users end up competing for air time. The trunked mobile system allows all available channels to be shared among a large group of users, so that no one is waiting as long as free channels are available. For instance, a typical trunked land mobile radio system can have up to 28 channels, one of which is a dedicated control channel that automatically assigns the open channels to users. When a user wants to transmit, both the transmitting radio and the receiving radios are automatically assigned to one of the 28 available radio channels. This system is better at conserving spectrum, since no individual radio user will have to wait to transmit if a trunked radio channel is available. Depending on how the available spectrum is used, these systems can also be classified as narrowband or wideband. In the narrowband architecture, the total frequency band is split into several narrowband channels. In the wideband architecture, most (or all) of the spectrum is available to each and every user.

Land mobile radio can be roughly divided into two separate classes of service: public and private. Paging, cellular telephony, and personal communications services (PCSs) fall under the classification of public land mobile radio, although some small paging networks may be private. We discuss each of these public service systems in turn in the next three sections. When cellular telephony is discussed, we also summarize the popular cordless telephony systems developed in Europe and Japan for comparison. We then turn to a discussion of private land mobile radio systems. Although some recent systems tend to support both private and public services, we discuss them together because many of them are evolved from early private systems. After discussing radio data networks, short-range beacons, and satellites communication technologies, we conclude this chapter with issues related to integration of communications subsystems. Information on the remainder of the technologies listed in Table 8.2 can be found in [4–7]. To save space, some system names, modulation methods, and open protocols contained in some tables are not spelled out, interested readers may contact the manufacturers for the full names of the systems or consult [2,8,9] for the full names of the technical terms listed.

### 8.3.1 Paging

Paging is the simplest mobile radio system. A basic paging system is shown in Figure 8.1, where PSTN stands for the public switched telephone network and BS stands for base station. It includes an input source, the existing wireline telephone

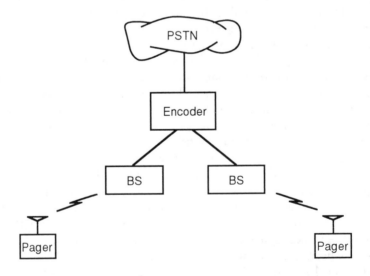

**Figure 8.1** Paging network.

network (PSTN), the paging encoder equipment (Encoder), the base station (BS) controlled by the transmitter control equipment, and the remote pager. Input can be entered from a phone, a computer with a modem, a desktop page-entry device, or through an operator who enters the message. Once the message is entered, it is sent via PSTN to the encoder. The encoder checks the validity of the pager number and looks in the subscriber database for the pager address. It then converts the address and message into the paging signaling protocol (a special code). This encoded paging signal is then sent through the control system to the transmitters located at each base station. The base station broadcasts the signal on the specified frequency for its coverage area to beep the pager.

Paging was originally developed as a one-way communications service (from a base station to mobile or fixed receivers) that provides signaling or information transfer by tone, tone-voice, numeric or alphanumeric message display, etc. Recent advances in technology have led to the introduction of two-way paging. This makes paging technology even more attractive as a medium for vehicle communications.

Radio paging has many attractive features, which explains why public interest in the technology has grown by leaps and bounds during the last decade. Paging systems are fairly robust and most can reach customers located deep inside buildings. The technology is widely available, and pagers are becoming even more versatile and portable as paging systems continue to take advantage of digital signaling technology. Competition among service providers is also creating a host of options for consumers, ranging from basic tone-only models (which beep or vibrate when someone is needed) to alphanumeric devices that can display short messages or other advanced models. For instance, tone-and-voice models will notify the user of a call and then relay a voice message at the customer's convenience.

One interesting paging method is to superimpose paging signals on AM/FM radio or TV signals. This signal multiplexing technology is generally referred to as *broadcast subcarrier*. We discuss it in further detail in Section 8.3.6. As far as applications of paging in vehicle location and navigation are concerned, for example, a transportation authority could establish its own paging network in outlying areas by contracting with FM radio stations to use FM subcarrier signals to transmit short traffic messages along with the programming broadcast. FM radio listeners would never hear these pages unless they tuned to that specific subcarrier frequency with a proper radio receiver. Using such an approach, the authority could establish a rural network without having to construct its own costly radio towers and transmitters. Such a paging system could serve as an adjunct to other commercial services and would come into use whenever mobile workers traveled beyond the range of the privately run networks.

In spite of recent advances, paging is still primarily a one-way communications network. Its strengths lie more in transmission of short data messages. Paging is restricted by two main shortcomings. First, it is primarily an urban service, although some paging networks reach further into the suburbs than do cellular systems. Second (and most critically), paging has traditionally been a one-way service. Users can receive brief messages, but they cannot acknowledge them unless they have a cellular phone or are near a wired telephone. Despite these drawbacks, paging is a highly affordable and widespread communications technology.

A multitude of paging companies and paging equipment is currently available on the market. For instance, Teletrac Inc. provides a paging service that operates in the 900-MHz band to provide two-way paging and tele-locator features. SkyTel's two-way paging service allows users to receive messages and respond with an acknowledgment via a button. This service takes advantage of wireless text messaging and adds the capability to send replies. Socket Wireless Messaging Service, or SWiMS, is a suite of services administered by Socket's National Dispatch Center and designed to turn an alphanumeric pager into a wireless link to the user's favorite communications resources: telephone, voice mail, fax, e-mail, modem, and paging system. Interested readers should consult the various trade magazines for wireless services based on radio paging technology and associated PCMCIA paging cards that are currently available. More information about various available paging equipment, devices, and protocols, can be found in [10,11].

### 8.3.2 Cellular

Cellular networks are full-duplex circuit-switched communication systems whose main components are of a set of mobile stations (MS, mobile or portable cellular phones), a base station (BS), a mobile switching center (MSC), and the public switched telephone network (PSTN) (Figure 8.2). The cellular system can provide both voice and data services over either analog or digital networks, although the main design objective of the analog network is to carry voice. Each coverage area in the network is divided into cells that are usually 1.5 to 30 km (0.93 to 18.65 miles) in diameter. The cell sites (BS) located at the center of each cell consist of a transmitter, receiver, control computer, and an antenna. The base station computer controls the frequencies, directs calls to open channels, and alerts the switching station when a call weakens. For simplicity, Figure 8.2 shows only two BSs associated with two cells. The switching station (MSC) connects the cellular phone to the local carrier and manages calls through frequency reuse, roaming, and hand-offs. Adjacent cells use different frequencies to avoid overlap, and cell site hand-offs occur automatically as each mobile station moves from cell to cell. A telephone carrier and its local office connects the call to the person being called and provides billing for the service.

The revolutionary concept behind cellular systems is the frequency reuse method proposed by Bell Telephone Laboratories. A certain frequency used by a channel in one area can be reused in another area. Each area consists of a cell. Cells using the same carrier frequency are located sufficiently far apart from one another such that interference can be within tolerance limits. As a result, each region initially

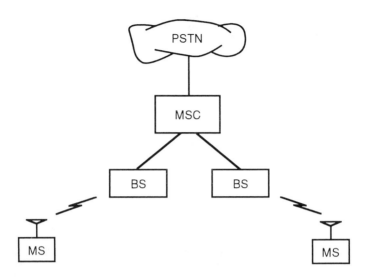

**Figure 8.2** Cellular network.

served by a single base station before introduction of cellular systems can be divided into several cells. Each of these cells will have its own base station and its own set of channels. Therefore, more services can be provided within the same frequency range.

Several different cellular radio systems are used in different areas around the world. The characteristics of these systems are summarized in Table 8.3 (where some items have been quoted from [12]). Many books are available that discuss the principles and applications of these systems so we do not discuss them further in this section. More information on these systems and the underlying principles may be found in [9,13–16]. Note that analog cellular systems use FDMA for channel access, while digital cellular systems use either TDMA or CDMA and can still provide analog services using FDMA if designed to do so. Digital cellular systems promise similar advantages over analog cellular systems mentioned in the beginning of Section 8.3. In Table 8.3, m-b is the broadcast rate from mobile station to base station, b-m is the broadcast rate from base station to mobile station, ISDN stands for Integrated Service Digital Network, and CEPT stands for the Committee of European Posts and Telegraphs.

Various techniques and ideas for using cellular systems to provide broadcast data for vehicle communications exist even though cellular systems were originally designed for one-to-one communications. On analog voice networks, off-the-shelf modems must be used to modulate circuit data over cellular links as briefly discussed in Sections 8.4 and 10.2.2. These services are good for mobile users who need to transmit or receive large amounts of data in a single session without worrying about

**Table 8.3**
Characteristics of Cellular Radio Systems

| Name | Date of Operation | Channel Spacing | Frequency (MHz) | Number of Channels | Characteristics |
|---|---|---|---|---|---|
| NAMT | 1978 | 25 kHz | 870–885 b-m 925–940 m-b | 600 | FDMA; increased to 1000 channels |
| NMT-450 | 1981 | 25 kHz | 453–457.5 m-b 463–467.5 b-m | 180 | FDMA; low channel capacity; good radio coverage; suitable for rural areas |
| AMPS | 1983 | 30 kHz | 825–845 m-b 870–890 b-m | 832 | FDMA; city-based; higher capacity than NMT, but smaller cells |
| C-450 | 1985 | 20/10 kHz | 451.3–455.74 m-b 461.3–465.74 b-m | 222/444 | FDMA |
| TACS plus ETACS | 1985 | 25 kHz | 890–915 m-b 935–960 b-m 872–888 m-b 917–933 b-m | 1000 plus 640 | FDMA; 50% greater capacity than AMPS, but smaller cells |
| NMT-900 | 1986 | 12.5 kHz | 890–915 m-b 935–960 b-m | 1999 | FDMA; designated for cities; caters to hand-held portables |
| NAMPS | 1992 | 10 kHz | 825–845 m-b 870–890 b-m | 2412 | FDMA; one 100-bps digital channel; 14-character mobile terminated messages |
| GSM | 1992 | 200 kHz | 890–915 m-b 935–960 b-m | 175 with 8 timeslots per channel (1400) | FDMA/TDMA; digital; ISDN capability; CEPT standard |
| TDMA (D-AMPS) | 1993 | 30 kHz | 825–845 m-b 870–890 b-m | 832 with 3 users per channel (2496) | TDMA; digital; IS-54 standard |
| PDC | 1993 1994 | 50 kHz 25 kHz interleave | 824–849 b-m 940–956 m-b 1429–1453 m-b 1477–1501 b-m | 1600 with 3 users per channel (4800) | TDMA, digital; for Japan only |
| CDMA | 1995 | 1228 kHz | 825–845 m-b 870–890 b-m | 19–20 (could vary) | CDMA; digital; IS-95 standard |

the extra call setup time and call disconnect time. In addition, a network overlaying the existing analog cellular network has been developed to address packet data applications. We discuss this network (CDPD) in comparison with other radio data networks in Section 8.3.5. See Section 10.4 for applications of the analog AMPS network (Table 8.3) in vehicle communications.

For comparison with cellular technology, Table 8.4 lists various cordless technologies that have been introduced in Europe and Japan. These systems are considered to be second-generation technology, mainly on the basis of the fact that they use digital signals and the fact that there are numerous base stations outside the home that can operate with a large number of individual handsets. By contrast, first-generation cordless technology is based on analog signals and only operates within residences between unique base station and handset pairs. This first-generation technology has found common use in the United States. Because of the relatively low cost and mobility, the new cordless telephony systems have become an integral part of personal communications services, as discussed in the following section.

### 8.3.3 Personal Communications Services

Personal communications services (PCS) have been defined by the U.S. Federal Communications Commission (FCC) to cover a broad range of radio communications services that free individuals from the constraints of the wireline PSTN and enable them to communicate when they are away from their home or office telephone [17]. Some literature refers to this technology using the term "personal communications network" (PCN). The PCS concept was developed out of the recognition that both cellular and cordless telephones have severe limitations. Therefore, PCS is supposed to combine the advantages of each system along with other added features to meet the communications requirements of people on the move. PCS has been proposed as a multienvironment, multioperator, and multiservice type system (Figure 8.3). Many people classify wireless systems based on analog technology as

**Table 8.4**
Characteristics of Cordless Telephony Systems

| Name | Date of Operation | Channel Spacing | Frequency (MHz) | Number of Channels | Characteristics |
|------|-------------------|-----------------|-----------------|--------------------|-----------------|
| CT2 | 1989 | 100 kHz | 864–868 m-b<br>864–868 b-m | 40 | FDMA/TDMA; digital |
| DECT | 1993 | 1728 kHz | 1880–1900 m-b<br>1880–1900 b-m | 10 with 12 users per channel (120) | TDMA; digital |
| PHP | 1995 | 300 kHz | 1895–1907 m-b<br>1895–1907 b-m | 77 with 4 users per channel (308) | TDMA; digital |

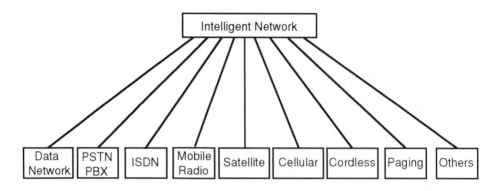

**Figure 8.3** Proposed PCS architecture.

first generation, systems based on digital technology as second generation, and PCS the next major step in the evolution of communication systems, as third generation [18].

Although there is no exact definition or well-defined system architecture for PCS, the basic concept is to support highly competitive wireless communications services without any restrictions on providers' capabilities and imaginations. PCS frequencies have been divided into two main groups by the FCC: wideband (broadband) frequencies and narrowband frequencies. The wideband PCS frequencies lie in the 1.8- to 2.2-GHz band; these frequencies are intended for high-throughput voice and data applications. Mobile devices (or handsets) might be able to roam freely among homes, offices, and roads, even access the Internet, using devices evolved from cellular phones. Narrowband PCS frequencies have been allocated to the 900-MHz band; these frequencies are intended for short and burst voice and data applications. These mobile devices might be small, using devices evolved from two-way pagers, and other personal computing or communications devices. Both wideband and narrowband PCS should be able to support vehicle location and navigation.

Some believe that PCS can be thought of as a family or continuum of wireless communication services at lower power and with more (and smaller) cell sites than cellular systems. Instead of using large transmitter towers, PCS is envisioned as employing transmitter/receivers small enough to be mounted inconspicuously on the sides of buildings and on top of telephone poles. This results in much smaller transmission cells (often called *microcells*). The minimum PCS cell size is expected to be approximately 0.8 to 2.5 km (0.497 to 1.554 miles) in radius. Consequently, PCS will require a larger number of cells for the same coverage area. This results in dramatically lower power requirements (but probably worse interference between adjacent cell sites than occurs for cellular services). Because PCS devices will use lower power, the phones should be smaller, lighter, and able to run longer on a single charge, perhaps even for weeks at a time. PCS is expected to be able to accommodate many more callers than cellular networks. As a result, promoters of this technology expect the rates for air time to be cheaper. In contrast, others think that there are other alternative architectures and implementations available. Some services will have more features and will presumably be more costly.

A wide variety of alternate versions of PCS have been proposed. Although the first U.S. PCS has been in commercial operation since November 1995, several other proposed systems are either still in test trials or in the process of being implemented. Many digital service specifications (including different voice-coding techniques and signal compression schemes to maximize spectrum usage) are being considered by rival PCS. Around the world, PCS has been allowed frequencies between approximately 800 and 3000 MHz. Proposed PCS services have included both licensed and unlicensed frequency alternatives [19]. The licensed services proposed for use include sophisticated pocket telephones, cellular phones at 2 GHz, and wireless fixed and short-distance links connecting the phones to microcells. Unlicensed services proposed for use include cordless business and home telephone systems, wireless local-area networks for computers, and wireless portable information devices. In short, PCS has the potential to improve mobile communications technologies existing today.

### 8.3.4 Private Land Mobile Radio Systems

As mentioned earlier, paging, cellular telephony, and PCSs are generally regarded as public land mobile radio services. In contrast, private land mobile radio services can be divided mainly into the categories of conventional, trunked, and specialized mobile radio (SMR). Although most SMR operations technically would fall under public services, the decision simply follows the private sector classification in the FCC frequency regulations (FCC: 47 CFR, Part 90). In the conventional radio system, users must share a common RF channel and compete for air time by monitoring the system for an open channel as described at the beginning of Section 8.3. Trunked radios are controlled by microprocessors, which act as automatic switchboards. Trunked radio system shares repeater(s), antennas, and channels. SMR systems were introduced to the United States in 1974 by the FCC to provide private land mobile services on a commercial basis. A SMR system can be either conventional or trunked, although most SMR systems are now trunked. These three types of systems can take different forms, such as individually owned, shared, voice only, data only, and integrated voice and data.

Although classified roughly under the title of this section, some recent systems intend to serve both the private and public sectors, such as the iDEN system, which is discussed later, and the European and Japanese systems listed later in Table 8.6. Because of these recent developments, the differences between private and public land mobile radio systems are becoming smaller, which has led to a blurring of the distinctions between these two categories. Unlike cellular technology, it is difficult to find public references discussing certain recent systems such as iDEN. Therefore, we cover this system in this section. General and in-depth descriptions of various land mobile radio systems can be found in [20,21].

Traditionally, private land mobile radio activity has been in the areas shown in Table 8.5 (after [1]). There are many different types of vehicle applications. One example of private data systems for vehicle location applications is the Motorola Private DataTAC system, which has been used in the United States and has a gross data rate of 9.6/19.2 Kbps. Another example is the system used for the dynamic route guidance system discussed in Chapter 11. Most radio equipment currently in service uses analog voice and would require modems to modulate, transmit, and receive position and traffic data over the air. As new systems are implemented, many are moving to digital technology.

Many land mobile radio systems combine voice and data via a trunked radio system. Several such trunked radio systems are shown in Table 8.6. The last four are mainly used in Europe (MPT-1327 and TETRA) and Japan (MCA and DMCA). The remainder of the systems are available in American and other markets.

Trunked radios have different implementations. A private trunked system can consist of a single site or a large multiple-site system. Each site has multiple repeaters that can receive, amplify, and retransmit radio signals. One repeater at each site is a control channel, which transmits a continuous stream of data (digital codes). Mobile and portable radios communicate with the site over the control channel to request an available channel. In a trunking system, each radio is assigned an individual unit ID and is also programmed by the system administrator with one or more groups that are user selectable. These groups, usually referred to as *talkgroups*, are

### Table 8.5
#### Traditional Private Land Mobile Radio Service Categories

| Land Transportation | Public Safety | Special Emergency | Industrial |
|---|---|---|---|
| Auto emergency | Fire | Beach patrols | Business |
| Motor carrier | Forest conservation | Communications standby facilities | Forest products |
| Railroad | Highway maintenance | Disaster relief | Manufacturers |
| School buses | Local government | Emergency repair of public communications facilities | Motion picture |
| Taxicab | Police | Handicapped | Petroleum |
| | State and federal governments | Medical services | Relay press |
| | | Rescue services | Special industrial |
| | | Veterinarians | Telephone maintenance |
| | | | Utility business |

**Table 8.6**
Representative Trunked Voice and Data Radio Systems

| Company | System | Data Protocol | Modulation | Channel Access | Gross Data Rate | Voice | Channel Spacing |
|---|---|---|---|---|---|---|---|
| Ericsson | EDACS | RDI | Binary FSK | FDMA | 9.6 Kbps | Analog/ digital (Aegis) | 25 kHz |
| E. F. Johnson | LTR | LTR | FSK | FDMA | 0.3 Kbps | Analog | 25 kHz |
| E. F. Johnson | Multi-Net | Multi-Net | FSK | FDMA | 9.6 Kbps | Analog | 25 kHz |
| Motorola | iDEN | LAPi | M-QAM | TDMA | 64 Kbps | Digital | 25 kHz |
| Open | APCO Project 25 | CAI, IP | QPSK-C | FDMA | 9.6 Kbps | Digital | 12.5 kHz |
| Motorola | ASTRO | CAI, IP, FLM | QPSK-C | FDMA | 9.6 Kbps | Digital | 25 kHz 12.5 kHz |
| Motorola | SmartZone SmartNet II | MDLC MDC-4800 RD-LAP, .. | FSK | FDMA | 1.8/3 Kbps 4.8 Kbps 9.6/ 19.2, ... | Analog | 12.5/25 kHz |
| Open | MPT-1327 | MPT-1327 | FFSK | FDMA | 1.2 Kbps | Analog | 12.5/25 kHz |
| Open | TETRA | TETRA | $\pi/4$ DQPSK | TDMA | 36 Kbps | Digital | 25 kHz |
| Open | MCA | MCA | MSK | FDMA | 1.2 Kbps | Analog | 12.5/25 kHz |
| Open | DMCA | RCR Standard-32 | M-QAM | TDMA/ TDM | 64 Kbps | Digital | 25 kHz |

groups of radios that will transmit and receive each other's communications. When a radio is set to a particular talkgroup, that radio will only be able to transmit and receive information from other radios in the same talkgroup. With talkgroups, each radio user will only hear the communications desired. One typical method for segmenting users into groups on a trunking system is according to functional responsibilities.

Requests for telephone interconnects are also handled through the control channel in similar fashion. A repeater is linked to a phone line and connected to the PSTN. This enables a radio system user to communicate with persons not on the radio system via standard telephone. Depending on the radio system, the conversation will either be in full duplex or half-duplex. Full duplex operates like a typical phone call, whereas half-duplex implies that the radio can only transmit or receive but not both simultaneously.

Some systems use a different approach. Instead of using a separate control channel to communicate with the site, they utilize slower and low-speed data on the voice channels. Although the trunking operation is similar, some features such as "busy queuing and callback" are not available in a system design of that type.

As mentioned earlier, SMR systems were introduced in the United States in the early 1970s, although significant amounts of equipment and systems did not begin to appear until the early 1980s. Trunked SMR sites can be linked, and also include data to provide a nationwide voice and data communications network. In the early 1990s, new radio systems (enhanced SMR or ESMR) were developed to compete with and exceed the services offered by cellular radio carriers. Licenses of many independent private SMR service providers were purchased and used together to provide more capacity and services on a single large network that integrates voice dispatch, wireless phone, text messaging (paging), and data transmission capabilities into one handset with cellular-like coverage. Unlike cellular systems, these ESMR operators have agreed to a single standard—the one set by the iDEN system discussed next.

### Integrated Dispatch Enhanced Network

The integrated Dispatch Enhanced Network (iDEN) is a digital communications system that combines the capabilities of a standard dispatch system with that of a cellular-like interconnect system. iDEN was developed by Motorola and is now in use in Argentina, Canada, Israel, Japan, the United States, and several other countries. iDEN uses an advanced proprietary modulation technology consisting of a speech compression scheme enabling six or three communication paths to be carried over a single 25-kHz RF channel. This technology is a radical change in the basic modulation techniques and hardware platforms that have dominated mobile radio for the past 40 years.

A key limiting factor for expansion of wireless communications products is global spectrum availability. By allowing six communication paths over a single channel, the iDEN system leads to a sixfold increase in efficiency. Furthermore, iDEN provides numerous other benefits including regional networking, full-duplex telephone interconnect, and integrated service. iDEN gives SMR operators the ability to integrate wireless phone, text messaging, two-way radio, and data transmission into one network using a single subscriber unit, referred to as a mobile station (MS, either mobile or portable). iDEN eliminates the need to carry a phone, a pager, a two-way radio, and a modem to stay in touch.

The digital technology used in iDEN provides additional benefits beyond integration of various services, including the following: more consistent voice audio quality throughout more of the coverage area, fewer dropped calls, improved security, and better system access.

Several other additional features of iDEN are listed below:

- Smaller hand-held portables;
- Full-duplex service (like cellular);
- Messaging services similar to digital paging that will allow rapid response;
- Superior data services to avoid lengthy verbal message exchanges and provide office mobility;
- Reduction of busy waiting time due to more efficient use of the RF spectrum;
- Advanced mobile telephone services;
- Advanced dispatch services (group communication with transparent networking);
- Flexible network scaling.

This technology allows fleets of users equipped with MS units to communicate with one another or the PSTN. The RF system consists of strategically located enhanced base transceiver systems (EBTSs), which are linked to base site controllers (BSCs) that direct the paths to either a dispatch application processor (DAP) for dispatch-type call processing or to a mobile switching center (MSC, or Northern Telecom Switch) for telephone interconnection (Figure 8.4). The BSC manages the RF channel, and the MSC maintains control over which users can enter the network, what their operating privileges should be, and where they are presently located within the system. Information previously entered into the home location register (HLR) database is used by the MSC for this maintenance. The EBTS site consists of one or more base radios (BRs), an RF distribution system, a BR frequency reference, a site synchronization receiver and antenna, a local-area network interface, and BR antennas. The EBTS provides the radio communications link between the land network and MS units. The BRs perform the communications with the MS units, sending both control information and compressed speech over a multiplexed radio channel. The EBTS can be configured to support multiple RF frequencies in an omnidirectional or sectored configuration.

In Figure 8.4, VLR stands for the visitor location register, which contains and is updated with current information such as most recent location, and the feature provisioning table. AUC stands for the authentication center, which holds the authentication keys for all MS users. MPS stands for the metro packet switch, which is a collection of components providing wide-area communications services. These services include distribution of incoming packets and dynamic replication of voice packets for transport to the respective BSC and corresponding EBTS. The switch provides connectivity between iDEN sites during a dispatch call.

The gross radio-channel bit rate for an iDEN system is approximately 64 Kbps. Most earlier systems utilized FSK (frequency shift key) modulation. In contrast, the iDEN system uses M16-QAM (quadrature amplitude modulation). This linear modulation provides a highly desirable combination of modulation efficiency, channel sensitivity, acceptable carrier-to-interference ratio trade-offs, and low adjacent channel interference.

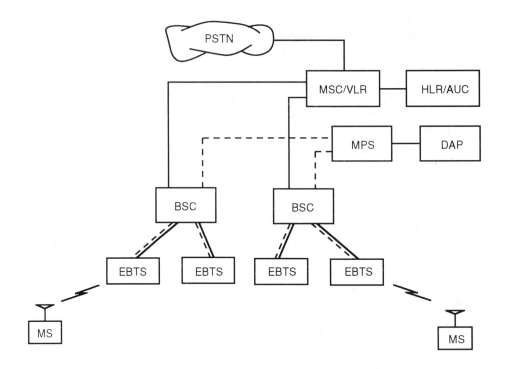

**Figure 8.4** Major elements of an iDEN system.

The iDEN equipment uses frequencies of 806-821/851-866 MHz, with a carrier spacing of 25 kHz. The separation between transmitting/receiving pairs is 45 MHz. The ability to accommodate 12.5-kHz offsets is also included. iDEN equipment may also be configured to operate in higher frequency ranges, such as 1453-1465/1501-1513 MHz with 48-MHz transmitting/receiving pair separation.

The compression algorithm used for digitally encoding/decoding speech is known as Vector Sum Excited Linear Predicting (VSELP). VSELP is a member of a class of speech coders called Code Excited Linear Predictive Coding (CELP) and Stochastic Coding (SC) or Vector Excited Speech Coding (VESC). It is the vocoder used in iDEN.

The iDEN implementation divides the RF carrier into discrete time slots of 15 ms (including overhead for transmitter turn-on, training and synchronization, and propagation delay). Auxiliary data embedded within each slot provide for associated signaling in addition to the normal digitized voice and data traffic. A set of slots on one group of RF carriers is devoted to trunking control. Inbound control-channel access is obtained via a reservation protocol. TDMA is the channel access technology used by the iDEN system.

iDEN allows an MS user to roam freely to any connected system and maintain full telephone interconnectivity (including circuit-switched data) and short messages, provided such services have been contracted for with the originating service provider. iDEN circuit-switched data are designed to be compatible with the following applications:

1. Laptop and palmtop computers;
2. Fax and image processing;
3. File transfer and other "dial-up" applications;
4. One-to-one connections between an in-vehicle user device and the PSTN.

iDEN circuit-switched data are compatible with the Radio Link Protocol (RLP) and full-duplex packet stream. In addition, MS users have the ability to receive short alphanumeric messages on their MS displays. A short message is generated by the Short Message Service-Service Center (SMS-SC). High-speed packet data service is being planned. Furthermore, iDEN provides system control with an authentication mechanism, which may be invoked prior to any chargeable service initiation. Clearly, iDEN can be used for placing various vehicle communication applications on a single bill since it is provided by a single nationwide service.

### 8.3.5 Radio Data Networks

The operational principle of radio data networks (RDNs) is similar to that of cellular networks except that RDNs are dedicated to data and are not circuit switched. RDNs consist of land-based radio base stations and efficient switching networks for packet data. Cellular-like frequency reuse and overlapping cells allow users to roam seamlessly. The network communication protocols either conform to the ISO Open Systems Interconnection (OSI) Reference Model (shown later in Figure 8.10) or the Internet protocols, which enable efficient interconnection and data transmission among all components of the network.

The growth of RDNs has been limited by the availability of unassigned radio spectrum. Although other widespread networks are in the planning phases, only two major RDNs are currently operating in the United States: ARDIS and RAM Mobile Data. These networks have been allocated frequencies in the 800- and 900-MHz bands and provide wireless packet data services to a large number of subscribers. Recently, several CDPD networks have been deployed and extended to selected cities where circuit-switched data services are used to send the packets to the CDPD network. Table 8.7 provides an overview of these and other networks. The Ricochet RDN shown in Table 8.7 was developed by Metricom, Inc., and uses the unlicensed 902- to 928-MHz band. It claims to be faster than the ARDIS and RAM Mobile Data networks. Further information on this and another proposed network can be found in [1,22].

**Table 8.7**
Radio Data Networks

| Network | Gross Data Rate | User Data Rate | Frequency | Data Protocol | Modulation | Channel Access | Channel Spacing | Modem |
|---|---|---|---|---|---|---|---|---|
| ARDIS | 4.8 or 19.2 Kbps | 2.4 or 9.6 Kbps | 800-MHz band | MDC-4800 RD-LAP | FSK | FDMA/ DSMA | 25 kHz | ARDIS, IBM, Motorola |
| RAM (Mobitex) | 8 Kbps | 4 Kbps | 900-MHz band | MASC IP | GMSK bt=0.3 | FDMA/ S-ALOHA | 12.5 kHz | Ericsson, IBM, RIM, Megahertz, Motorola, etc. |
| CDPD | 19.2 Kbps | 9.6 Kbps | 800-MHz band | TCP/IP PPP/SLIP | GMSK | FDMA/ DSMA | 30 kHz | Cincinnati Microwave, SierraWireless, PCSI, etc. |
| Ricochet | 76 Kbps | 19.2 Kbps | 902-928 MHz | TCP/IP | GMSK | FH SS (ISM) | 160 kHz | Metricom, Inc. |

## ARDIS

One of the first RDNs to be established in the United States, ARDIS grew out of a private wireless data network that Motorola had built to support IBM's nationwide staff of field technicians. In 1990 the network made its services publicly available to business firms needing to stay in touch with mobile workers. ARDIS now has more than 60,000 end-users in 400 metropolitan areas and 10,700 cities and towns across the United States. This network is typically used to support two-way data transfers of less than 10K in size. The maximum packet size is 240 bytes (not including the 16-bit address field). Remote users can access the network via hand-held or laptop computers with radio modems that communicate with the base stations. Each base station is connected to one of 32 radio network controllers managed by two ARDIS hosts. These hosts, as access points, are linked to the customers' mainframe computers via leased telephone lines using asynchronous, SNA LU6.2, or X.25 dedicated circuits.

The ARDIS network architecture is cellular-like with overlapping cells. The base station provides line-of-sight coverage up to a radius of 16.09 to 24.14 km (10 to 15 miles). ARDIS can support portable communications from inside buildings and on the street because of the overlapping coverage, designed power levels, and error-correction coding in the transmission format. ARDIS has between one and three duplex pairs in each of its service areas. Each duplex channel has two 25-kHz channels spaced 45 MHz apart. ARDIS supports data transmission rates of up to 4.8 Kbps using its equipment (MDC-4800) and 19.2 Kbps on upgraded equipment (RD-LAP), for a maximum user data rate of about 8 Kbps.

Both host routing and peer-to-peer messaging options are available. Host routing is used when a wireless device sends messages to and receives messages from a fixed host connected to the ARDIS network. This method of routing is suitable for applications that require central control functionality or interoperability with other networks and fixed systems and is usually accomplished using a dedicated private line operating under the X.25 PTP protocol. Peer-to-peer routing supports exchange of messages between individual wireless devices. These messages are routed to the destination wireless devices by the ARDIS network.

## RAM Mobile Data

The RAM Mobile Data network is based on the Mobitex system developed by Ericsson in Sweden. RAM operates the Mobitex network throughout the United States, covering more than 7,700 municipalities, including airports and major transportation corridors. More than 40,000 subscribers currently use RAM's service for applications including field transportation, field service, field sales, finance, utilities, telemetry, point-of-sale transactions, public safety, and messaging.

The RAM network is hierarchical in structure, with the network control center, which manages the entire network, forming the top layer. National switches, which route traffic between service regions, form the next layer (immediately below the network control center). The next two layers are occupied by regional switches and local switches, each of which handles traffic within a given service area. The base stations, which communicate with the mobile and portable data devices, are at the bottom. The host computers are connected at the local switch layer using standards such as X.25, TCP/IP, SNA LU6.2, and SNA 3270, with speeds of up to 64 Kbps. RAM provides an interconnected trunked radio network that is configured like a cellular phone system, except that hand-offs are managed by the mobile terminal itself rather than the network.

The RAM network has 10 to 30 two-way radio channels with a bandwidth of 12.5 kHz in each metropolitan service area. Its hierarchical architecture enables RAM to route messages via the shortest available path for increased speed and reliability. Data are sent between devices in the field and the base stations over the air at 8 Kbps in half-duplex mode with a maximum file size of 20K and a user data rate of 4 Kbps. The maximum packet size is 512 bytes (not including the 24-bit address field).

## Cellular Digital Packet Data

Cellular digital packet data (CDPD) is a fast, efficient, digital data transmission system that overlays existing analog cellular networks. The system operates at 19.2 Kbps per channel and internetworks with existing data networks based on ISO 8473 (CLNP) or the Internet Protocol. CDPD provides maximum connectivity by using idle times between cellular voice calls or by transmitting in a dedicated-channel environment.

Some people believe that CDPD should be gaining prominence over other land-mobile-radio data technology because of the following. CDPD is designed to be reliable and secure. CDPD's data transmission technology is efficient, using the bandwidth not utilized by voice. CDPD utilizes much of the cellular infrastructure currently in place for the voice network, allowing for quick implementation and reduced network costs. CDPD's open network design approach permits technological evolution and service extensions. Recently, the U.S. Joint Architecture Team for national ITS selected CDPD as the preferred network for providing traffic and route guidance information to motorists [23].

CDPD operates on a network that consists of the basic components shown in Figure 8.5. Mobile End Systems (M-ESs) are portable subscriber devices that permit the user to gain access to the network by sending and receiving CDPD radio signals. Mobile data base stations (MDBSs) are located within each cell site and act as a data relay between the M-ES and the MD-IS (see remainder of paragraph). The

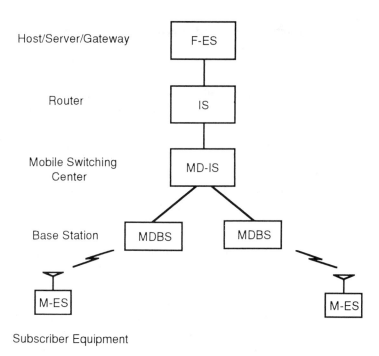

**Figure 8.5** CDPD network.

MDBS communicates with the MD-IS via a 56-Kbps data link. The Mobile Data Intermediate System (MD-IS) performs routing functions based on its knowledge of which M-ESs are in its service area. The Intermediate System (IS) acts as a router, which completes the connection between the MD-ISs and the Fixed-End Systems (F-ESs) by making complex routing decisions and supporting data links as they change over time. The F-ES is equivalent to the traditional data application or internal network support and service application. Its location is geographically fixed. A F-ES is equivalent to a host, server, or gateway that can connect to public data networks, corporate customer networks, or the Internet.

The primary new technological components introduced by the CDPD network involve the use of existing cellular channels for air data link services and mobility management. The use of existing cellular channels for air data link services required introduction of the CDPD Airlink Interface. This specification defines all procedures and protocols necessary to allow efficient use of cellular channels for data communications.

Mobility management satisfies the requirement to provide continuous communications to mobile data subscribers. The CDPD System Specification defines procedures and protocols necessary to ensure correct and efficient delivery of connectionless data packets between end systems. Connectionless network service enables the CDPD network to provide simple, reliable cell-transfer mechanisms. End-to-end connections are maintained by the mobile system and the application service provider.

With more than 16,000 AMPS cell sites existing in North America, CDPD has better potential nationwide subscriber coverage than competing packet-data wide-area networks, provided special CDPD equipment is available. It also has a high active user capacity. However, CDPD has some weaknesses. There is additional infrastructure cost to run voice and data side by side. Voice is given priority over data and may cause blocking of data services. The idle channel quality and foreign carrier detect features will cause CDPD to interfere with voice systems. To date, CDPD has not been widely deployed—it has seen much less than ARDIS in the United States. Only time will tell if CDPD can overtake ARDIS and RAM. Refer to [24] for more information on CDPD.

In conclusion, both ARDIS and RAM were developed for applications involving short wireless messages. Currently, Ricochet has limited mobility support so most of its applications use wireline modems. Ricochet may not be suitable for mission-critical applications because it uses unlicensed bands and must yield to licensed users whenever there is interference. At present, it has less coverage than either ARIDS or RAM. Finally, CDPD has the potential to provide the largest coverage. The recently developed Circuit-Switched CDPD Specification enhances the coverage by allowing mobile devices in areas of no CDPD coverage to dial in and access a CDPD MD-IS. CDPD is suitable for applications requiring wireless short messages, burst messages, or lengthy sessions that can afford sporadic communications.

### 8.3.6 Broadcast Subcarriers

Broadcast subcarriers can provide voice or data communications from a broadcast station or infrastructure to vehicles without requiring a specially allocated frequency spectrum. Multiplexed subcarrier technology was first introduced in the 1930s when FM was invented by Armstrong [25]. The very name *subcarrier* implies that it is not the main signal carrier generated by a broadcast station. Instead, it is a radio signal (voice or data) broadcast along with the main signal on a sideband frequency assigned to a specific station because the main signal does not occupy the entire allocated spectrum. Although this technology was developed concurrently with FM, subcarrier signals can also be broadcast with amplitude modulation (AM) and television signals. In early days, any such signal transmission required a Subsidiary Communications Authorization (SCA). Even though the SCA was phased out in 1983, people still refer to these signals as SCAs. Traditionally, subcarriers have been used for a variety of subscription services: background music, weather forecasts, sports events, stock quotations, foreign-language soundtracks, special time signals, or other information relevant to business, professional, educational, or religious interests, etc. Special receivers are required to listen to these signals. Recently, subcarriers have become very popular for broadcasting traffic information.

There are two broadcasting technologies. In monophonic broadcasting, only one audio signal is transmitted. Only one speaker is required to reproduce this type of signal. In stereophonic broadcasting, two audio signals are combined and transmitted over a single channel. A stereophonic receiver can separate the two signals to reproduce the original signals via separate speakers. A monophonic receiver may also receive and reproduce the stereophonic signals, but the spatial distribution effect will be lost.

As mentioned earlier, since all broadcast stations (AM, FM, and TV) have the ability to transmit subcarriers, they provide an effective solution for broadcasting traffic information. Subcarriers can transmit both audio and digital information side by side on the same channel in such a way that listeners with conventional receivers cannot hear the digital transmissions on the data subcarrier. Special receivers are required to access the data part of the channel. Table 8.8 shows some of the technical performance parameters of various subcarrier options (from the FCC: 47 CFR 73.319, [26,27]). AM subcarrier broadcasting is rarely used because of the relatively slow speed (10 to 12.5 bps), but higher data rates (up to a few hundred bits per second) are possible [28]. Note that in Europe and Asia (regions 1 and 3) the frequency allocated for each AM channel is only ±9 kHz around the main carrier instead of ±10 kHz (20 kHz total) for the United States (region 2).

As an example, FM radio stations can support a standard background music subcarrier audio (SCA signal) in addition to an analog audio channel with a frequency response from 50 Hz to 5 kHz. Since there is room for two such channels on a

**Table 8.8**
Technical Performance Characteristics of Subcarriers

| Station Type | Subcarrier Bandwidth | Injection | Main Channel Loss |
|---|---|---|---|
| AM monophonic | None (20 kHz)* | None | None |
| AM stereophonic | 0–12.5 Hz (25 Hz) (20 kHz)* | 5%, −32 dBc, ±2.86° | 0 dB |
| FM monophonic | 79 kHz (20–99 kHz) | 23.5 kHz | 2 dB |
| FM stereophonic | 46 kHz (53–99 kHz) | 15 kHz | 1 dB |
| TV monophonic | 104 kHz (16–120 kHz) | 50 kHz | 0 dB |
| TV stereophonic | 73 kHz (47–120 kHz) | 25 kHz | 0 dB |

*Total bandwidth occupied by the transmitted composite signal.

stereophonic FM carrier—one centered at 67 kHz and the other at 92 kHz, two data channels can be supported. Each channel can support up to 4.8 Kbps of data throughput without any impact on the station's audio quality. Several systems attempt to take advantage of this data capacity, such as the Secondary Audio Program (SAP) and PROfessional channel (PRO), the Vertical Blanking Interval (VBI) system, DARC (which was developed by NHK), HSDS (which was developed by Seiko), STIC (which was developed by MITRE), and the RDS/RBDS standards (which are discussed below). These systems are all listed in Table 8.9.

*Radio Data System*

The radio data system (RDS) was first defined by the European Broadcasting Union (EBU) and various European broadcasters, notably Swedish Telecom Radio (STR) and the BBC. Development of this system has been facilitated International Radio Consultative Committee (CCIR) Recommendation 643. Car radios that implement RDS first became available in 1987, and CENELEC (European Electrotechnical Standards Organization) is developing a standard for RDS receivers. The main purpose of RDS is to facilitate automated tuning by using data labels to identify the data segments being broadcast to RDS receivers. RDS uses an inaudible data-modulated subcarrier, added to the stereo multiplex signal from conventional FM broadcast transmitters. One of the first data-modulated subcarrier systems of this type was the MBS (mobile search) paging service, which was developed in Sweden by STR. MBS provided the initial basis for the system that became the EBU's RDS, and there are many similarities in the basic data transport system. From about 1976, development work on RDS was coordinated by a group of technical experts working under the auspices of the EBU. The specification was unanimously approved by all EBU members in 1983 and was published as EBU Technical Document 3244 in 1984. The following high-level requirements were established:

1. RDS signals must be downwardly compatible; they should not cause interference to reception of main program audio signals on existing receivers or

**Table 8.9**
Broadcast Subcarrier Systems

| Systems | Station | Status | Interface | Gross Data Rate | User Data Rate |
|---------|---------|--------|-----------|-----------------|----------------|
| DARC | FM | Operational | Proprietary | 19 Kbps | 8 Kbps |
| HSDS | FM | Operational | Proprietary | 19 Kbps | 7.5 Kbps |
| RDS/RBDS | FM | Operational | Open | 1.187 Kbps | ≈0.3 Kbps |
| SAP & PRO | TV | Operational | Open | 0.3–9.6 Kbps | 0.3–9.6 Kbps |
| STIC | FM | Concept | Proprietary | 18.8 Kbps | 7.6 Kbps |
| VBI | TV | Operational | Open | 1.2–19.2 Kbps | System dependent |

interfere with the operation of receivers which use the ARI (Autofahrer Rund-funk Information) system (assist in tuning and alert the driver of imminent broadcasting).

2. The data signals should be capable of being reliably received over a coverage area at least as great as that of the main program signal.

3. The usable data rate provided by the channel should support the basic require-ments of station and program identification, and provide sufficient scope for future development.

4. The message format should be flexible to allow the message content to be tailored to meet the needs of individual broadcasters at any given time.

5. The system should be capable of being reliably received on low-cost receivers.

Under the EBU RDS specification, a broadcaster can choose a subcarrier level such that the deviation relative to the subcarrier frequency is between 1 and 7.5 kHz, with a 2-kHz deviation recommended for most circumstances. A 57-kHz subcarrier is generally chosen (Figure 8.6) because it is a harmonic of the 19-kHz pilot tone. This subcarrier is suppressed and amplitude modulated by the shaped and biphase-coded data signal.

The RDS signal is transmitted as a biphase-coded signal with a data rate of 1187.5 bps. Along with the specified 100% cosine roll-off filtering, this provides an overall bandwidth of approximately 4.8 kHz for the data signal. The signal consists of 16 possible data groups as listed in Table 8.10. Like most serial data transmission systems, RDS partitions each data group into blocks: four blocks for

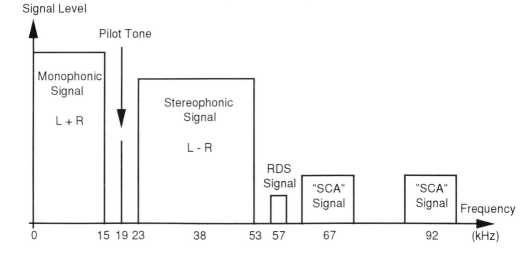

**Figure 8.6** Spectrum of a FM stereophonic multiplex signal with RDS and SCAs.

**Table 8.10**
RDS Data Groups

| Group Type | Functional Description |
| --- | --- |
| 0 | Basic tuning and switching |
| 1 | Program item number |
| 2 | Radio text |
| 3 | Location and navigation |
| 4 | Clock time and data |
| 5 | Transparent data channels |
| 6 | In-house applications |
| 7 | Radio paging |
| 8 | Traffic message channel (TMC) |
| 9 | Emergency warning |
| 10 | Program type name |
| 11 | Undefined |
| 12 | Undefined |
| 13 | Undefined |
| 14 | Other enhanced networks |
| 15 | Fast tuning and switching |

the longest (104-bit) group. Each block has a 16-bit data word and a 10-bit checkword formed with a cyclic code shortened by burst error correction. The 10-bit checkwords are modified by various offset words to form a self-synchronizing code. On the frame receiving side, the decoder uses this self-synchronizing code to identify the beginning and end of the received data [29]. In other words, the checkwords in the burst-error-correcting code are used to indicate correct block synchronization and address the contents of each block. This code enables the decoder to reliably acquire synchronization within a one block period. The decoder obtains both block and stream synchronization using the different offset words contained in each of the four blocks that make up each group. The group type is embedded in the first 4 bits of the second block in each group. The fifth bit indicates the RDS version number, either A (value 0) or B (value 1).

A mixture of either fixed or variable RDS groups may be transmitted (in any order). The group format can be carried over from one minute to the next in order to meet different requirements at various times of day. The purpose of this mixture of fixed and variable format is to provide rapid access to tuning data while retaining the flexibility to allocate capacity to various applications to meet the different needs of individual broadcasters at any given time. If the required capacity is available, new group types can be placed into the transmitted data stream without affecting the operation of existing RDS receivers.

Although traffic information is very popular with European drivers, there is evidence that too many messages detract from satisfaction and make it very difficult for broadcasters to handle. This service has been transmitted silently on FM under group type 8A (see Table 8.10—note that A is the version number), although it is not yet widely used. Other information, including emergency, weather, and radio programming information, can be displayed in textual format. In an effort to standardize the service provided by the traffic message channel (TMC), European countries have attempted to define the RDS-TMC service more completely and guide its implementation. More information about this coordinated effort can be found in [30].

*Radio Broadcast Data System*

The Radio Broadcast Data System (RBDS) is an outgrowth of RDS, and was developed by the National Radio Systems Committee of the United States under the sponsorship of the Electronics Industry Association and the National Association of Broadcasters. During development of the standard, an attempt was made to keep the RBDS standard compatible with RDS. However, a number of modifications had to be made because of the different structure of the broadcasting industry in the United States. RBDS turns out to cover both AM and FM broadcasting (while the European version only covers FM broadcasting). Although RBDS is a modification of RDS, it fully includes RDS. Therefore, the operational principles and data groups described in the last section are all valid under RBDS as well.

There are few only differences between RBDS and RDS [31]. RBDS includes an option for multiplexing RDS with the slightly modified MBS developed by Cue Paging. It also includes an option to use an in-receiver database permitting some sort of RDS functionality for AM and FM stations that do not implement RDS. RBDS has an option to add a yet to be defined AM data system. A few new concepts have been added, such as group type 3A being reserved for differential GPS (DGPS, Section 3.4.2), channels 0 and 1 of group type 5 being reserved for the in-receiver database system updates, and channel 2 being reserved as an SCA switch. The DGPS pseudorange correction data are sent by the reference stations using the RTCM SC-104 (Radio Technical Commission for Maritime Services Special Committee 104) industry-wide standard.

As we see from this section, RDS/RBDS technology can provide traffic, emergency, and weather broadcast features (in addition to news and entertainment) at the touch of a button, provided the user uses an RDS/RBDS receiver (radio). This explains why this broadcast subcarrier technology has become very popular in Europe and why it is now available for dynamic navigation systems in Japan (VICS, Section 10.3.2). The RDS and RBDS standards can be found in [32] and [33], respectively. More information about other broadcast systems using various subcarrier technologies can be found in [23,27,34].

### 8.3.7 Short-Range Beacons

Short-range beacons are a way of supporting vehicle-to-roadside communications. Beacons provide short-range communications and can transfer data at high speeds using a limited spectrum. Depending on the design, beacons can be used for one-way periodical broadcasting, two-way broadcasting and receiving, or two-way point-to-point communications. Besides vehicle location and navigation, beacons can also be used in many other applications such as electronic (automatic) toll collection, automatic vehicle identification, commercial vehicle operations (CVO), traffic management, and vehicle-to-vehicle communications. Several recent examples involving the application of this technology to vehicle location and navigation can be found in Section 10.3.

Either microwave or infrared beacons can be used. In Europe and Japan, microwave beacons operate in the 2.5- and 5.8-GHz bands [35–38], and infrared beacons operate at wavelengths of 850 and 950 nm. CEPT has assigned space in the 63-GHz band for development of vehicle-to-vehicle communications. In the United States, although the 915-MHz band is more popular in products, the FHWA is preparing a petition to the FCC for allocating 5.8-GHz band for ITS applications. Table 8.11 provides a comparison of three beacon technologies available in the United States (after [23]).

There are three types of vehicle-to-roadside beacons: (1) location beacons, (2) information beacons, and (3) individual communications beacons [39]. Location beacons transmit signals identifying their location, map coordinates, road segment heading, and beacon number. Information beacons transmit both location signals and relay current road and traffic information received via cable. Individual communication beacons are used for two-way communications with vehicles. These beacons can be used to collect traffic data and guide the vehicle.

As it passes a communication beacon, an appropriately equipped vehicle can transmit measured travel times and waiting times experienced at traffic lights via the beacon to a centralized host. Meanwhile, it can receive relevant location and guidance information back from the beacon. For example (see Figure 8.7), each

Table 8.11
Short-Range Beacon Characteristics

| Manufacturer | System | Range | Date Rate | Transmit Block Size |
|---|---|---|---|---|
| Hughes | Active RF | 61m (200 ft) | 550 Kbps | 512 bits |
| Amtech | Passive RF | 23–30m (75–100 ft) | 300 or 600 Kbps | 128 bits |
| Siemens | Infrared | 60–80m (197–262 ft) | 125 Kbps | 256 bytes (downlink) 128 bytes (uplink) |

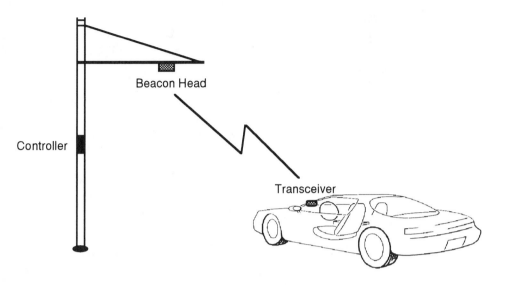

**Figure 8.7** Beacon communication.

infrared beacon site in the EURO-SCOUT system (described in Section 10.3.1) consists of a beacon controller and up to 16 beacon heads that can be installed on road facilities such as sign posts, traffic lights, and poles [37]. Depending on the applications, beacon controllers and heads could be installed at every intersection, at every fourth traffic light, etc.

Figure 8.8 shows the relationship between the number of infrared beacon heads required, the transmission data rate, and the vehicle travel speed. These results are based on transmission data rates of 500 Kbps (downlink) and 125 Kbps (uplink) with a BER of $10^{-7}$. One interesting observation is that the increase in the size of the region covered is proportional to the square root of $n$, where $n$ is the number of beacon heads.

Beacon systems typically use dedicated short-range communication (DSRC) protocols. The main organizations working on DSRC standards are the European Committee for Standardization (CEN) incorporating with the ISO [40] and the American Society for Testing and Materials (ASTM) [41]. Similar to the digital map database standard, there is still no universally agreed upon DSRC standard available. Although international compatibility among beacon systems remains limited, many agree that the new standards must support multiple applications, be open to further evolution, be developed within a short time frame, ensure privacy and security, and be independent of such communications media as microwave and infrared.

One asynchronous protocol proposed in Europe and one synchronous protocol used in North America may be used as examples. The asynchronous protocol is

Number of Beacon Heads

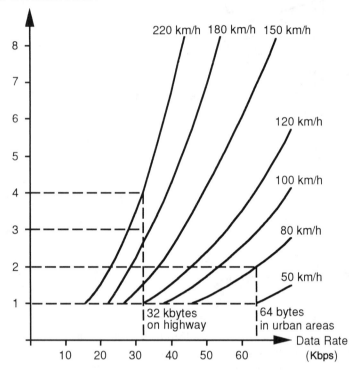

**Figure 8.8** Required beacon heads under various transmission data rates and travel speeds.

based on a half-duplex TDMA approach [42]. Each beacon periodically broadcasts a flexible sequence of messages on the downlink and uplink. The downlink contains the Beacon Service Table (BST) and other data, and the uplink contains public and private windows (messages) being transmitted to the vehicle. Under the asynchronous protocol, when a vehicle enters the coverage zone of a beacon, it is able to receive the BST, which contains a list of services offered, along with operational parameters. The vehicle communicates with the beacon by accessing a public window which always follows the BST. Once the vehicle ID has been transmitted to the beacon using the public window, a set of private windows is reserved for further data exchange for this vehicle. If a beacon or vehicle transmission fails, the beacon will retransmit the downlink window and offer an uplink window once the failed window has expired. If no transmission occurs during one of the steps in the sequence, the beacon assumes that the vehicular unit has experienced technical difficulties or has left the coverage zone. The beacon then reverts to its initial state, ready to transmit the BST again.

The synchronous protocol (developed by Hughes) is a fixed-frame structure based on TDMA (Table 8.11). It consists of a reader control message (equivalent to the BST message), activation slots (equivalent to the public windows), and slot data messages (equivalent to the private windows). Detailed information can be found in the draft standard developed by the ASTM to describe a similar protocol [41].

One relatively old technique is also used to provide vehicle-to-roadside communications: the loop detector (or proximity beacon). A loop detector is a magnetic inductive sensor embedded in the road surface to detect traffic volumes and occupancies or to communicate with the vehicle [43]. Loops are typically deployed in a rectangular or circular (for the newer loops) configuration, with one loop per lane. The electronics associated with the loop are generally installed in a control cabinet, together with other intersection control equipment. A recent advance in traffic-flow monitoring enhances the output accuracy of a standard loop detector for counting passing vehicles to only one error in 10,000 [44].

Compared with cellular systems and RDS/RBDS, beacon communications offer high transmission rates, effective position calibration, location-oriented traffic information, and the ability to accurately detect and measure the parameters of vehicles on a specific road or in a specific lane. The disadvantages are that the communication zones are very limited, communication is discontinuous, and system installation and maintenance costs are high.

### 8.3.8 Satellites

Satellite communications systems consist of a space segment and a ground segment. The space segment consists of the satellite and control station where all the operations associated with monitoring the vital functions of the satellite are performed [45,46]. The ground segment consists of all Earth stations that use the operational satellite. As shown in Figure 8.9, these Earth stations may either transmit or receive.

Satellites can be classified into two broad types: geosynchronous (geostationary or GEO) satellites and low-Earth-orbit (LEO) satellites. GEO satellites orbit approximately 41,326 km (22,300 nautical miles) above the equator. In synchronization with the Earth's rotation, these satellites remain over approximately the same locations on the Earth's surface (near the equator). Because they are far above the earth, relatively large antennas are required. GEO satellites have dominated the field as the technology of choice for many years. Recently LEO satellites have been showing popularity, although most of these systems are still in development. LEO satellites orbit much lower in the sky, typically about 500 to 1500 km (310 to 932 miles) above the ground [with some systems around 10,000 km (6200 miles) above the ground]. Unlike GEO satellites, LEO satellites are not restricted to equatorial orbits. Some of them have circular orbits while others have elliptical orbits. There are little LEO satellites that operate at VHF frequencies (or slightly higher) and carry only data

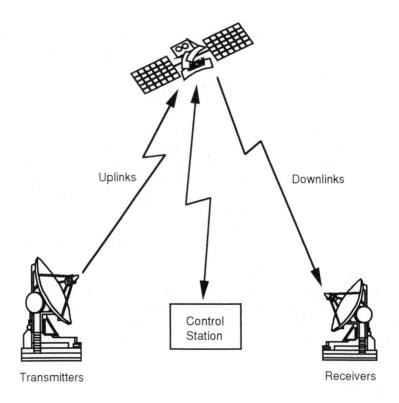

**Figure 8.9** A satellite-based communications network.

communications, and big LEO satellites that operate above 1 GHz and carry both voice and data communications (VHF: between 30 and 300 MHz). Each category of satellite has its own advantages and disadvantages.

Despite these differences, all satellites and satellite operations have certain features in common. All communications satellites have transponders. The transponders pointed toward Earth are used to receive signals from Earth stations. They then relay these signals to distant receiving stations on the ground. The former type of transmission is called an *uplink* and the latter is called a *downlink* (Figure 8.8). Some of the early satellites had only one transponder on board, but it is now common for modern satellites to carry a few dozen transponders.

When satellites receive radio signals from ground-based antennas, they cannot simply relay those same signals to a transponder pointing at the ground-based receiving station. Otherwise, this would imply that the two transponders were using the same carrier frequency, which would lead to interference between the uplink and downlink. To avoid the interference problem, the satellite translates the received frequency to a different frequency. It then amplifies the signal for transmission to the ground, since the signal was weakened during uplink. Although this technique reduces interference between uplink and downlink, additional frequency allocations are required.

Technology has also been developed to control the shape and size of the region on the Earth's surface to which the satellite broadcasts. This area of signal reception is known as the *footprint*, which is a circular or shaped beam from space to the reception points on Earth. Satellite downlinks often have broad footprints for easy reception. To avoid offending certain regions or countries with unwanted signals, a technology for obtaining smaller, better controlled and condensed footprints has been developed. This has enabled the reuse of limited frequencies among satellites for cellular-type networks.

To prevent mutual interference, satellite slots have been established every 2 degrees along the equatorial orbital plane, which means that only 180 slots are available in geosynchronous òrbit; few of these slots are still unoccupied. As the number of satellites in orbit has increased, so has the number of transponders aboard each satellite. More frequency bands have been required in order to support this space-based technology. The most common bands for satellite transmissions are VHF, L-band (1-GHz band), C-band (the 3-, 4-, 5-, and 6-GHz bands), and Ku-band (the 11-, 12-, and 14-GHz bands). Table 8.12 lists several existing satellite systems that offer voice, data, and/or position services.

Due to the growing need for mobile communications, new satellite systems will continue to be deployed. Three big LEO systems and three little LEO systems have recently been approved by the FCC in the United States. One big LEO system has also been approved for its frequencies by the World Radio Communications (WRC) Conference. As mentioned in the beginning of this section, the orbits of some systems (Odyssey and OCI) are higher. They are sometimes called medium-Earth-orbit (MEO) satellites or intermediate circular orbit (ICO) satellites In Europe. One little LEO system, Orbcomm, is already in operation (Table 8.12). The remainder of the planned systems are briefly described in Table 8.13. Readers must bear in mind that the information collected here is believed correct as of this writing. However, it may be changed or updated anytime during the system development phase.

Satellite technology has not been widely accepted for many surface vehicle communications (except the heavy trucking industry) although it has been available since the early 1970s. The main reasons are that the mobile terminals are large and equipment and service costs are high. To attract more users, the size and weight of mobile terminals must be further reduced, and affordable equipment and services must be available while still providing efficient transmission.

## 8.4 COMMUNICATIONS SUBSYSTEM INTEGRATION

Once the appropriate communications technology has been selected based on the attributes discussed in Section 8.2 and other factors, the equipment must be integrated

**Table 8.12**
Existing Satellite Systems

| Satellite System | Organization | Service | Date of Operation | Coverage | Terminal Types | Voice Coding Rate | Gross Data Rate | Modulation | Channel Access | Voice Bandwidth |
|---|---|---|---|---|---|---|---|---|---|---|
| INMARSAT-A, B, M, Mini-M (GEO) | INMARSAT | Voice, data, fax | 1976, 1993, 1993, 1996 | Global, excluding polar regions | Marine vessels, large portables, small portables | N/A, 16, 4.8, 4.8 Kbps | 9.6, 9.6, 2.4–4.8, 4.8 Kbps | FM, QPSK, QPSK, QPSK | Voice: FDMA Data: TDMA and TDM | 50, 20, 10, 5 kHz |
| INMARSAT-C (GEO) | INMARSAT | Telex, e-mail, position | 1990 | Same as above | Land mobile, maritime and aeronautical portables | N/A | 0.6 Kbps (store and forward) | BPSK | TDMA and TDM | N/A |
| Satellite Paging (GEO) | SpaceCom Systems | Paging | 1986 | North America | Many types of paging terminals | N/A | 1.2–512 Kbps | FSK | FDMA/TDMA | N/A |
| OmniTRACS (GEO) | Qualcomm | Data, position | 1988 | Global | Qualcomm terminals | N/A | 0.384 Kbps | MSK | SDMA/TDMA/CDMA | N/A |
| AMSC-1 [Skycell] (GEO) | American Mobile Satellite | Voice, data, position | 1995 | North America | Mobile data terminals (Mitsubishi, Trimble, Westinghouse, etc.) | 4.2 Kbps | 6.4 Kbps | QPSK | TDMA | 7.5 kHz |
| Orbcomm (Little LEO) | Orbcomm | Data, GPS data | 1995 | Global | Land mobile, fleet, cellular data, SMR data, GPS, many applications | N/A | 2.4 Kbps uplink; 4.8 Kbps downlink | PSK | FDMA | N/A |

*Note:* N/A = not applicable.

**Table 8.13**
Planned LEO Satellite Systems

| Satellite System | Organization | Service | System Status | Coverage | Subscriber Equipment | Voice Coding Rate | Data Rate | Modulation | Channel Access |
|---|---|---|---|---|---|---|---|---|---|
| Globalstar (Big LEO) | Globalstar, L. P. (Loral, Qualcomm, etc.) | Voice, data, fax, paging, GPS | FCC license Jan. 1995; Commercial operation projected in 1998 | Global | Qualcomm design | 9.6 Kbps variable | 1.2–9.6 Kbps | QPSK | CDMA |
| Iridium (Big LEO) | Iridium, Inc. (Motorola, Raytheon, etc.) | Voice, data, fax, paging, position | FCC license Jan. 1995; Commercial operation projected in 1998 | Global | Motorola, Scientific Atlanta | 2.4/4.8 Kbps | 2.4 Kbps | QPSK | FDMA/TDMA |
| Odyssey (Big LEO) | Odyssey Telecom. International, Inc. (TRW, Teleglobe) | Voice, fax, PC data, short messages | FCC license Jan. 1995; Commercial operation projected in 2000 | Global | Magellan, Panasonic, Nortel, JPC, Mitsubishi | 4.2 Kbps | 2.4–9.6 Kbps | QPSK | CDMA |
| ICO (Big LEO) | ICO Global Communications (U. K.) | Voice, data, fax, paging | WRC approved frequencies Nov. 1995; Commercial operation projected in 2000 | Global | Ericsson, Nokia, NEC | 4.8 Kbps | 7.2–9.6 Kbps | QPSK | TDMA |
| GE Starsys (Little LEO) | GE Americom, Inc. CLS (France) | Data, position | FCC license Nov. 1995; Commercial operation projected in 1998 | North America, Europe initially | TBD | N/A | 0.6 Kbps | MSK | DS-SSMA |
| VITASAT (Little LEO) | VITA | E-mail | FCC license July 1995; Operation projected in late 1997 | Global | TBD | N/A | 9.6 Kbps | GMSK | FDMA |

*Note:* N/A = not applicable.

with applications at both ends of the wireless network (mobile end and host end). For the most part, each interface is unique to the communications technology selected and will likely be proprietary to the equipment manufacturer. In some cases, however, the equipment interface follows an industry standard and off-the-shelf interfaces may be available. The current trend in equipment interfaces is moving toward the Internet suite of protocols on both the mobile and host ends. This greatly simplifies the task of interfacing new communications equipment at all levels. In addition, an open architecture may be very helpful, especially in terms of allowing users' "plug and play" usage of many different communications devices.

Actual implementation of the network interface protocols may depend on the type of host computing equipment, hardware interface, and operating system or application being used. For example, the host computer may be a mainframe computer, a UNIX workstation, or a personal computer (PC) that might be interfaced to the wireless network via dial-up lines, dedicated phone lines, or other wireline infrastructure. Similar implementation choices must be made on the mobile side. Most implementations on both host and mobile ends conceptually follow the ISO OSI Reference Model as shown in Figure 8.10, except that some layers in the OSI reference model may be added, replaced, or combined in the actual implementation. Each layer in the model provides a specified set of services to the layer immediately above and uses the services provided by the layer immediately below it to transport the message units [47]. The function of each layer is defined as a protocol describing

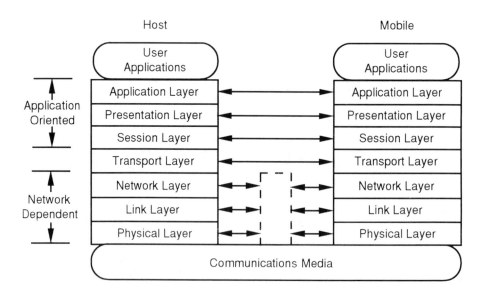

**Figure 8.10** ISO OSI Reference Model.

the set of rules and conventions used by that particular layer to communicate with the other layers.

On the system level, the quality of network service must be considered. For data services in particular, the considerations include acceptable error and loss level, average and maximum delay, and average and minimum throughput. For vehicle communications based on existing communications networks, more attention must be given to the transport layer, since the protocol used there can be chosen to optimize use of the underlying network-dependent layers. For integration, it is recommended that a standard set of protocols be used in each of the required layers, although customization will still generally be necessary. For instance, use of the popular TCP/IP (Transmission Control Protocol/Internet Protocol) will make the transport layer much easier to interface with others.

For RF communications, generally speaking, the higher the transmission rate, the higher the BER. Because mobile radios are constantly moving, the operational environment is much more severe than for radios used in a stationary environment. Error detection and correction become an important issue when ensuring reliability of data services. Furthermore, many vehicle location and navigation systems (Chapter 10) are now starting to use existing cellular networks as their communications support. New protocols and standards have become available for cellular data modems. Many different types of modems are now available on the market. However, their performance and ease of use vary. In addition, modem selection is affected by transmission data size and other factors. Therefore, a thorough study should be conducted before committing to one choice.

One possible integration method is to use the stream networking interface standard on either the host or mobile side (when UNIX-derivative operating systems such as pSOS+ or VRTX are used). This standard offers a solution to protocol proliferation problems. Streams were originally developed for UNIX-based network applications. They provide standards for protocol development, interfaces between protocol layers, and interfaces between networks and applications. Streams define a standard set of services for developers to implement protocol stacks that are independent of operating system kernel. This independence ensures portability across different operating environments. Off-the-shelf protocols can be easily integrated into applications containing stream-compatible protocol interfaces. Applications can be written so that they are independent of the underlying protocols and physical communication networks. Because of the popularity of streams, a great deal of off-the-shelf networking protocol software and device-driver software is available for streams. Many existing communication protocols are compatible with streams, which gives users access to a wide variety of networking software. Stream packages for embedded applications in a real-time environment have also been developed. They are much more compact and flexible than the original UNIX version.

The transport layer interface (TLI) is a widely used interface for applications in a stream environment. It can be applied to connection-oriented protocols such as ATM (asynchronous transfer mode) as well as X.25 and connectionless network protocols such as Ethernet and token ring. TLI can dynamically control or reconfigure stream device drivers, buffers, and timers. It can handle network data in packets or frames.

Similar strategies are available for the DOS and MS-Windows environments. Interfaces for both the mobile and host computer can be written at the DOS device driver level, to support a number of different network standards including the open data-link interface (ODI), packet-driver, and network device interface specification (NDIS), as well as the Virtual device Driver (VxD) and dynamically linked library (DLL) interfaces. Although most interfaces are evolving toward the MS-Windows VxD driver standard, NDIS drivers are portable across environments. One example is that Microsoft Windows NT provides TCP/IP host interfacing via its built-in network components. In this context, serial line internet protocol (SLIP) and point-to-point protocol (PPP) interfaces required by CDPD devices are simple to integrate. Other wireless interface requirements can also be simplified for the application once the underlying communications software has been written to work within the NDIS framework. This task requires a good understanding of the underlying network architecture.

To conclude this chapter, consider all of the communications technologies discussed in this chapter. Some technologies are restricted to one-to-one communications and some are restricted to one-to-many (broadcast) communications. Can you determine one by one which of these technologies are restricted, which technologies can be efficiently configured for one-to-one communication, and which can be configured for one-to-many communication? Furthermore, which technologies are suitable for one-way communication and which are suitable for two-way communications? Similar questions can be raised for many of the other system attributes discussed in this chapter, such as coverage, capacity, and cost. Understanding the main communications capability of each technology should enable us to select the proper network for any particular applications.

## References

[1] S. D. Elliott and D. J. Dailey, *Wireless Communications for Intelligent Transportation Systems*, Norwood, MA: Artech House, 1995.

[2] J. D. Gibson (Ed.), *The Mobile Communications Handbook*, Boca Raton, FL: CRC Press, 1996.

[3] M. D. Yacoub, *Foundations of Mobile Radio Engineering*, Boca Raton, FL: CRC Press, 1993.

[4] A. Santamaria and F.J. Lopez-Hernandez (Eds.), *Wireless LAN Systems*, Norwood, MA: Artech House, 1994.

[5] J. Z. Schanker, *Meteor Burst Communications*, Norwood, MA: Artech House, 1990.

[6] D. L. Schilling (Ed.), *Meteor Burst Communications: Theory and Practice*, New York: Wiley, 1993.

[7] W. H. W. Tuttlebee (Ed.), *Cordless Telecommunications in Europe: The Evolution of Personal Communications*, Berlin: Springer-Verlag, 1990.

[8] E. A. Lee and D. G. Messerschmitt, *Digital Communicaiton*, 2nd ed., Boston: Kluwer Academic, 1994.

[9] T. S. Rappaport, *Wireless Communications: Principles and Practice*, Upper Saddle River, NJ: Prentice Hall, 1996.

[10] J. Moskowitz, *Paging Protocols*, http://village.ios.com/~braddye/protocol.html, 1996.

[11] P. Seah, *An Introduction to Paging: What It Is and How It Works*, 2nd ed., Singapore: Motorola, Inc., 1994.

[12] J. Walker and B. R. Gardner, "Cellula Radio," in *Mobile Information Systems*, J. Walker (Ed.), Norwood, MA: Artech House, 1990, pp. 59–103.

[13] W. C. Y. Lee, *Mobile Cellular Telecommunications: Analog and Digital Systems*, 2nd ed., New York: McGraw-Hill, 1995.

[14] M. Mouly and M. B. Pautet, *The GSM System for Mobile Communications*, Mouly and Pautet, 1992.

[15] R. Prasad, *CDMA for Wireless Personal Communications*, Norwood, MA: Artech House, 1996.

[16] S. Redl, M. Weber, and M. W. Oliphant, *An Introduction to GSM*, Norwood, MA: Artech House, 1995.

[17] Federal Communications Commission, *Notice of Inquiry*, General Docket 90–314, 5 FCC Record 3995, 1990.

[18] V. O. K. Li and X. Qiu, "Personal Communication Systems (PCS)," *Proc. IEEE*, Vol. 83, No. 9, Sep. 1995, pp. 1210–1243.

[19] S. J. Lipoff, "Personal Communications Networks Bridging the Gap Between Cellular and Cordless Phones," *Proc. IEEE*, Vol. 82, No. 4, Apr. 1994, pp. 564–571.

[20] G. C. Hess, *Land-Mobile Radio System Engineering*, Norwood, MA: Artech House, 1993.

[21] E. N. Singer, *Land Mobile Radio Systems*, 2nd ed., Englewood Cliffs, NJ: Prentice Hall, 1994.

[22] L. H. M. Jandrell, Communication System and Method for Determining the Location of a Transponder Unit, *United States Patent No. 5365516*, Nov. 1994.

[23] Joint Architecture Team, *ITS Architecture: Communications Documents*, Federal Highway Administration, U.S. Dept. of Transportation, June 1996.

[24] M. Sreetharan and R. Kumar, *Cellular Digital Packet Data*, Norwood, MA: Artech House, 1996.

[25] E. H. Armstrong, "A Method of Reducing Disturbances in Radio Signaling by a System of Frequency Modulation," *Proc. Inst. Radio Eng.*, Vol. 24, No. 5, May 1936, pp. 689–740.

[26] Motorola, Inc., *Notice of Proposed Rule Making*, FCC Docket No. 21313, Oct. 1978.

[27] E. Small, "Broadcast Subcarriers for IVHS: An Introduction," *Proc. IVHS America 1993 Annual Meeting*, Apr. 1993, pp. 158–165.

[28] P. Dambacher, *Digital Broadcasting*, London: The Institution of Electrical Engineers, 1996.

[29] S. R. Ely and D. J. Jeffery, "Traffic Information Broadcasting and RDS," in *Mobile Information Systems*, J. Walker (Ed.), Norwood, MA: Artech House, 1990, pp. 141–175.

[30] D. Bowerman, "Pan-European RDS-TMC Traffic Information Services (DEFI Initiative)," *Proc. Second World Congress on Intelligent Transport Systems*, Nov. 1995, pp. 2566–2571.

[31] D. Kopitz and T. Beale, "The Radio Data System 'RDS' in Europe and the Radio Broadcast Data system 'RBDS' in the USA—What Are the Differences and How Can Receivers Cope with both Systems?" *RDS Forum*, Jan. 5, 1993.

[32] CENELEC, *Specification of the Radio Data System*, CENELEC EN 50067, Brussels, Belgium/Geneva, Switzerland: CENELEC and EN, 1992.

[33] NRSC, *United States RBDS Standard*, Washington, D.C.: Electronic Industries Association and National Association of Broadcasters, Jan. 1993.

[34] B. L. Hinton, H. W. Lam, and J. W. Marshall, "Design Trade-Offs for the Subcarriers Traffic Information Channel Waveform," *IVHS J.*, Vol. 1, No. 4, 1994, pp. 329–344.

[35] M. Ando and K. Takeuchi, "Infrared Vehicle Detector (IRVD) System Design," *Proc. Second World Congress on Intelligent Transport Systems*, Nov. 1995, pp. 1543–1548.

[36] S. Baranowski, M. Lienard, and P. Degauque, "Beacon-Vehicle Link in the 1-10 GHz Frequency Range," in *Automotive Sensory Systems*, C. O. Nwagboso (Ed.), New York: Chapman & Hall, 1993, pp. 271–291.

[37]  H. Sodeikat, "Dynamic Route Guidance and Driver Information Services with Infrared Beacon Communication," *Proc. Second World Congress on Intelligent Transport Systems*, Nov. 1995, pp. 622–627.

[38]  J. Takahashi, H. Hatashita, T. Nagashima, K. Abe, and S. Horii, "A Radio Beacon Receiver for Vehicle Information and Communication System," *Proc. Second World Congress on Intelligent Transport Systems*, Nov. 1995, pp. 1525–1529.

[39]  T. Saito, J. Shima, H. Kanemitsu, and Y. Tanaka, "Automobile Navigation System Using Beacon Information," *Proc. IEEE-IEE Vehicle Navigation and Information Systems Conference (VNIS '89)*, Oct. 1989, pp. 139–145.

[40]  European Committee for Standardization (CEN), *Road Traffic and Transport Telematic (RTTT)— Dedicated Short-Range Communication (DSRC): Physical Layer Using Microwave at 5.8 GHz*, Brussels: CEN, prENV 12253, Dec. 1995.

[41]  ASTM, *Standard for Dedicated, Short Range, Two-Way Vehicle to Roadside Communications Equipment*, Philadelphia, PA: American Society for Testing and Materials, Draft, Sep. 1995.

[42]  C.H. Rokitansky and C. Wietfeld, "Performance Evaluation of Data Link Layer (MAC/LLC) Protocols Proposed for Standardization," *Proc. First World Congress on Applications of Transport Telematics and Intelligent Vehicle-Highway Systems*, Nov./Dec. 1994, pp. 396–405.

[43]  R. L. Anderson, "Electromagnetic Loop Vehicle Detectors," *IEEE Trans. Vehicular Technology*, Feb. 1970, pp. 23–30.

[44]  S. Dunstan and B. Lees, "IDRIS and Loop-Based Tolling," *Traffic Technology International '96*, 1996, pp. 294–297.

[45]  G. Maral and M. Bousquet, *Satellite Communications Systems: Systems, Techniques and Technology*, 2nd ed., Translated by J. C. C. Nelson, New York: John Wiley & Sons, 1993.

[46]  D. Roddy, *Satellite Communications*, 2nd ed., New York: McGraw-Hill, 1996.

[47]  A. S. Tanenbaum, *Computer Networks*, 3rd ed., Upper Saddle River, NJ: Prentice Hall, 1996.

# PART II
▼▼▼

# SYSTEMS

Part II is devoted to a discussion of vehicle location and navigation systems. We shall learn how to use the modules studied in Part I to construct working systems. Vehicle location and navigation systems can be broadly categorized into two classes: autonomous and centralized. Whenever location and navigation functions are performed in the vehicle (mobile end) with no remote host or centralized computing facilities (host end) involved, we will refer to the system as *autonomous*. Otherwise, we will call it *centralized*. This classification is used for the convenience of our study. Different criteria may result in a different classification of systems discussed in this part.

Good system architecture is very important for successful location and navigation systems. Users' requests and developers' improvements generally lead to system expansion. A good architecture must provide a stable basis for the future evolution of the system. In other words, the system must be stable and predictable, but still flexible enough to meet changing demands and operational environments over a reasonable fraction of the expected system lifetime. On the other hand, unrestricted enhancement of a finished system (even when supported by a well-defined architecture to begin with) may affect system stability. Experience indicates that whenever a system is expanded by a factor of two or more, completely new problems will be encountered [1]. Once the architecture has been defined, it is often very expensive and difficult to change it later.

---

Several of the earlier chapters in this book are devoted to a discussion of the individual subsystems and technological building blocks for intelligent vehicle system architectures. We should always remember that when building a complicated integrated system, many of the architectural building blocks will have their own design specifications or objectives. These objectives may not always be consistent. We must harmonize these conflicting goals in order for the system integration phase to proceed successfully. Furthermore, integration of different modules is not simply a matter of data communication. Many other issues must be considered in addition to the basic system architecture. We will begin looking at these issues with a discussion of autonomous location and navigation systems in Chapter 9, a discussion of centralized location and navigation systems in Chapter 10, and a case study of an operational dynamic navigation system in Chapter 11.

### Reference

[1] Jesty, P. H., and J. Geizen, "It's Architecture Jim, But Not as We Know It," *Traffic Technology International*, June/July 1996, pp. 26–30.

# CHAPTER 9
▼▼▼

# AUTONOMOUS LOCATION AND NAVIGATION

## 9.1 INTRODUCTION

Autonomous location and navigation systems can be designed to various levels of complexity. System complexity is generally determined by architectural and design trade-offs. These trade-offs involve various architectural constraints such as system-level requirements for accuracy of location, the unit cost of the system, the complexity of the navigation functions to be supported, whether a wireless communications receiver is required to support the location function, and other considerations specific to the application.

Depending on the user requirements, an autonomous system could be designed to be as simple as a hand-held device or as complicated as a sophisticated in-vehicle navigation system. For example, let us consider the trade-offs associated with developing a low-cost, hand-held, autonomous location device used by recreational hikers or campers. We may assume that market-driven requirements will place a constraint on the cost of materials and production for such a hand-held unit. A simple implementation might be to incorporate a single inexpensive location determination technology, such as GPS, without additional subsystems to correct or remove certain position errors (i.e., DGPS). Thus we have traded lower unit costs for an order of magnitude reduction in position accuracy (most likely 100 versus 15 m) although some of the inaccuracy is intentionally imposed under selective availability (SA; Section 3.4.2). Additionally, given our cost constraints, we decide that a map display is also too expensive. As a result, the display unit to be used will provide a simple, text-only, black-and-white LED readout of the latitude and longitude of the device. Based on the display selection, we have made another significant trade-off. We have reduced system functionality by not including a map display in the unit. Note that the method of navigation used in this example is assumed to be "manual." Customers may need to plot the displayed latitude and longitude on a paper map (carried with them) to further confirm their location. A system of this minimal complexity may be suitable for recreational purposes such as hiking, cycling, or other low-technology transportation. Other, more sophisticated implementations require more architectural complexity.

A more complicated system implementation might use multiple integrated position determination technologies. An in-vehicle navigation system might use dead reckoning combined with GPS, a sophisticated map-matching capability, and a CD-ROM, hard drive, or memory card-based map database. The user interface may have a heads-up display (HUD) or a touch screen controlled color monitor capable of displaying both text and map graphics in conjunction with speech recognition and speech synthesis capabilities. It might also include route planning and turn-by-turn guidance for trip navigation. Some navigation components could be integrated with existing entertainment equipment already available in the vehicle.

In the remainder of this chapter, we consider both simple and complex implementations for autonomous location and navigation systems. We begin by discussing location determination methods for relatively simple location systems followed by a discussion of the methods used in more complicated navigation systems. Although high-precision location systems can be very complicated (in design and implementation), we use "simple" to indicate that such systems have fewer modules (as defined in this book) than navigation systems.

## 9.2  VEHICLE LOCATION

In this section, we discuss the location technologies used in mobile applications. A simple location architecture consists of two main modules (building blocks): a locator module and a user interface module as shown in Figure 9.1. As the figures indicates, various technologies are available. Depending on user requirements, we can select one from each module or mix different technologies to construct a working on-board location system.

Even though the architecture of this simple location system seems trivial, the location system is an essential and fundamental component of any advanced location and navigation system. We must pay special attention to the detailed design and implementation issues involved as discussed in Chapter 3. Without a solid location

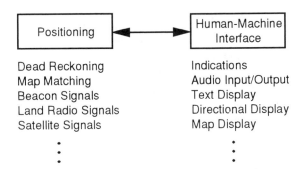

**Figure 9.1** Simple in-vehicle location system.

component, any complicated system built on top of it is doomed to have unsatisfactory performance. In the next three sections, we discuss three classes of location technologies that can be employed by a positioning module. For a rough comparison of these technologies, refer to Table 10.1.

### 9.2.1 Stand-Alone Technologies

The first type of location technology, stand-alone technology, is differentiated from the second and third in that it does not require a communications receiver to determine vehicle location. Recall from Section 3.2 that dead-reckoned positioning, a stand-alone technology, utilizes fusion of sensor data to determine the location of the vehicle. This technology depends on the system deriving an initial vehicle position, receiving a position, or retrieving a position (either entered by the user or stored at the time the vehicle last stopped). It then utilizes distance and directional information provided by on-board sensors to calculate relative coordinates in two-dimensional planar space. Relative distance measurements are usually derived from the vehicle odometer, and directional information is generally provided by a magnetic compass, differential odometer, or gyroscope.

   Accuracy may be an issue with systems of this type, due to errors introduced by noise at the sensor, or inaccurate determination of the initial vehicle position. Errors tend to accumulate over time, degrading the long-term position accuracy. Overall, systems requiring a moderate to high degree of location accuracy should not utilize dead-reckoning technology as the sole source for location determination. The advantage of this type of system is that it does not require a communications receiver for position measurement. This tends to reduce the overall cost of implementation. This reduced cost must, of course, be weighed against reduced accuracy. Over the long term, systems requiring more than a minimal level of location accuracy may eventually end up combining this technology with other location technologies.

As discussed in Chapter 4, map-matching algorithms can be used to resolve the inaccuracies caused by the dead-reckoning method. Various fusion technologies mentioned in Section 3.5 are also useful in compensating for the inaccuracy of dead reckoning. There are some alternatives that require external assistance. For example, short-range beacons could be installed on roadside signs or poles to transmit the known positions of these beacons to passing vehicles (Section 8.3.7). These data can then be used to calibrate the on-board positioning system. We will briefly discuss position determination and other applications of these beacons in Sections 10.2.1, 10.3.1, and Chapter 11. You may have already noticed that short-range beacons alone can be used for positioning. The limitation is that short-range beacons cannot locate vehicles traveling in between beacons. If more beacons are installed, this will increase the cost of the system. GPS, LORAN-C, and other technologies can also be used to improve the performance of stand-alone systems. However, once external assistance is being provided, the location technology becomes hybrid rather than stand-alone. These communications-receiver-based technologies are discussed in subsequent sections.

Another stand-alone location technology that does not require a communications receiver for location determination involves the use of a computer vision system. The basic principle of this technology is to store images obtained along the roads before the actual trip and then to match these prerecorded images with a series of images captured by cameras during the trip to identify the location of the vehicle [2]. A similar idea (and some variations on this idea) have also been used in other fields such as robotics [3]. One recent development has involved the integration of computer vision technology and dead reckoning [4]. This system combines the dead-reckoning module with a neural network trained to identify roadway landmarks encoded in a CD-ROM image database, which is correlated with a digital map database. Location determination is performed in real time, assisted by data compression and high-speed video capture techniques. The system is claimed to be accurate within approximately 5m. A block diagram of this system is shown in Figure 9.2.

As a stand-alone technology, this computer-vision-based method has both advantages and disadvantages. On the minus side, the image for the area to be traversed must have been recorded and stored in database form, which means that this technology cannot be used in areas not contained in the database. On the plus side, the system does not require use of a communications receiver for location determination. In addition, the pattern recognition and resultant determination of location is absolute in nature and offsets the detrimental effects of the position errors and sensor errors associated with dead reckoning alone. Although the method discussed uses natural landmarks for positioning, the principle can also be used with artificial landmarks. In other words, specially designed objects or markers could be placed along the road for the sole purpose of position determination, provided the extra installation and maintenance cost is not a concern. This would make the on-board system much simpler. This and other stand-alone technologies might, in a

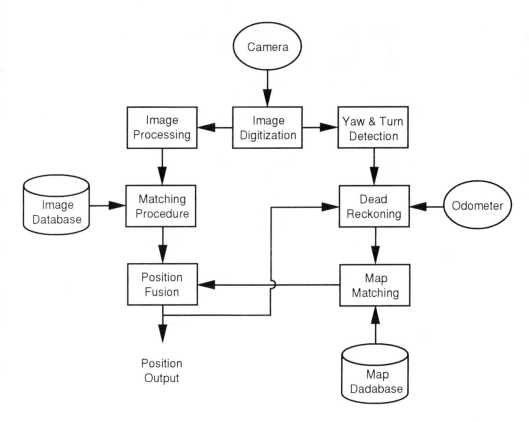

**Figure 9.2** A computer-vision-based automatic vehicle location (AVL) technology.

hybrid implementation, compensate well for a common deficiency often found in radio-based location technologies. Many radio location systems are subject to interference (multipath reflection/fading) from large objects such as buildings and hills. By contrast, computer-vision-based technology depends on these same large objects and may provide critical location data precisely where radio location systems lose accuracy. Besides automatic vehicle location, computer vision has been used in other vehicle applications such as automatic driving, automatic vehicle identification, collision avoidance, and automatic incident detection. These systems are beyond the scope of this book.

### 9.2.2 Terrestrial Radio Technologies

Before introducing actual location systems based on terrestrial radio signals, we first address the techniques commonly used for terrestrial radio-based location determination. Note that the methods discussed below can be used either at the mobile end or the host end located on a fixed infrastructure. The mobile equipment could be a stand-alone hand-held device. Although most of our discussions are vehicle centered, they can easily be generalized to hand-held devices. Based on the classification described in the beginning of Part II, if these methods determine the vehicle location at the mobile end, the corresponding system is called an autonomous system. If they determine the vehicle location by a host computer on the infrastructure end, it is called a centralized system.

Three commonly used measurement techniques for terrestrial positioning are time of arrival (TOA), angle of arrival (AOA), and time difference of arrival (TDOA). All of these approaches use the concept that the RF signal propagates at a constant velocity and that the signal path is predictable. The first technique, TOA, measures the propagation time of signals broadcast from multiple transmitters at known locations to determine the location of the mobile device or vehicle. This is the same technique used in GPS positioning. The difference is that for terrestrial positioning the emitters are not in space, but on the Earth's surface, typically taking the form of base stations or towers. A detailed discussion of the TOA principle may be found in Section 3.4.2.

The AOA technique uses RF triangulation to calculate the vehicle position. In infrastructure-based implementations, the signal is transmitted from a vehicle equipped with an RF transmitter. In this approach, a phased array of two or more antennas is used at a single cell site to receive the propagating wave. The following equation is often used for the two-antenna array at any one site to calculate the angle of incidence for a two-antenna array located at a single site, as shown in Figure 9.3.

$$\hat{\alpha} = \arcsin\frac{c\Delta t}{d}$$

where $\hat{\alpha}$ is the estimated angle of incidence of the propagating wave (assumed planar) at the antenna array, $c$ is the speed of light (assuming that the radio-wave velocity is approximately equal to the speed of light), $\Delta t$ is the difference between the times of arrival signal at each antenna, and $d$ is the distance between the antennas used to receive that signal. Note the assumption that the radio-wave velocity is approximately equal to the speed of light, which may not be valid in certain applications.

An interesting special case solution can be made by the observation that a single phase wave striking two closely spaced antennas at any one site will show a difference in electrical phase of the two received signals. Given that $d = 0.5\lambda$ (where $\lambda$ is a radio signal wavelength), the estimated incidence angle (arriving azimuth angle) becomes

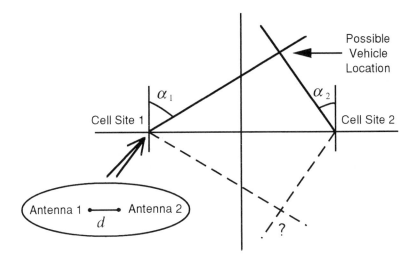

**Figure 9.3** Location determination by angle of arrival (AOA).

$$\hat{\alpha} = \arcsin\left(\frac{\Delta\phi}{\pi}\right) = \arcsin\left(\frac{\phi_1 - \phi_2}{\pi}\right)$$

where $\phi_1$ and $\phi_2$ are arriving electrical phase angles for antennas 1 and 2, respectively.

The phased arrays of the antennas are used at two or more cell sites capable of receiving the propagated signal from the vehicle. The location of the vehicle is determined by the intersection of the two angles of incidence $\alpha_1$ and $\alpha_2$ as shown in Figure 9.3.

We have shown an infrastructure-based approach in which each site has two antennas. A three-antenna array is actually better because a two-antenna array will have difficulty calculating the angle of incidence when it is close to a right angle. Theoretically, these antennas could instead be on the mobile side to receive the propagating radio waves from transmitters at the base stations. For economic reasons, antenna arrays are seldom used on the mobile end.

There are both advantages and disadvantages to using the AOA for the determination of vehicle position. On the positive side, there is no need to maintain time synchronization between cell sites (or base stations) to perform vehicle positioning. Only two sites are required to determine the location of a vehicle. Because it does not use a multisite time-synchronized system (as TDOA does), the overall performance of AOA as a location technology should be less affected by RF channel bandwidth. This is an important feature to keep in mind when dealing with various RF technologies [such as 30-kHz AMPS and 10-kHz NAMPS at the low end to 1.25-MHz CDMA at the upper end (Table 8.3)] in a single system.

The major drawback to AOA is its susceptibility to signal blockage and multipath reflection. This results in fairly high error margins for the estimated vehicle positions. This is especially true in urban areas, where errors on the order of hundreds of meters can occur. Because of signal scattering, it is conceivable that position calculations based on AOA could result in a position estimate that places the vehicle in the opposite direction from its actual direction relative to the receiving base sites (as indicated by the question mark in Figure 9.3). These ambiguous solutions can be eliminated using additional technologies such as the RF profile method discussed in Section 10.2.1. Another problem with AOA is that each site or mobile device (depending on infrastructure-based or mobile-based solution) needs to have at least two antennas (which adds additional cost to the system). However, this may not be a problem for some established sites that have a phased array of antennas already.

The third type of a terrestrial-radio-based location technique is TDOA. The TDOA technique utilizes RF trilateration to calculate the vehicle position. RF trilateration differs from triangulation in that it calculates the distance between the vehicle and a fixed set of reference sites that are time synchronized. The calculated distance from the vehicle can be determined by either of two methods, measuring the transit time for a radio signal (group of pulses) between the vehicle and reference sites, or the total phase change in the radio signal between the vehicle and reference sites. The method using pulse modulation for the radio signal is less affected by multipath propagation than the method using phase modulation, which means that pulse modulation is more accurate. On the other hand, pulse modulation requires a higher bandwidth than phase modulation. The radio signal could also be transmitted first from the site to the vehicle with the vehicle then responding back to the site. In this case, the calculated distance must be divided by two. Figure 9.4 illustrates the basic principle of this location technology.

We now discuss how TDOA technology determines a location. If we have a time-synchronized signal (either generated by the moving vehicle or by time-synchronized fixed RF transmitters) at sites 1, 2, and 3 as shown in Figure 9.4, we can determine the signal transmission path lengths $d_1$, $d_2$, and $d_3$. The differences between these path lengths can be measured by the time (or phase) differences of the signals between the transmitters and the vehicle. The estimated location derived from the time or phase difference pairs will be at the intersection of two hyperbolas as shown in Figure 9.4. Assuming that the time difference is used to derive the distance and a receiver in the vehicle is used to receive the signals transmitted from three sites, these hyperbolic curves may be calculated as follows (using the curve $h_1$ as an example): The mobile receiver detects the pair of transmissions from sites 1 and 2, and determines the difference in arrival times $\Delta t_{12}$. This time difference can be translated into a path length difference as follows:

$$d_1 - d_2 = c\Delta t_{12}$$

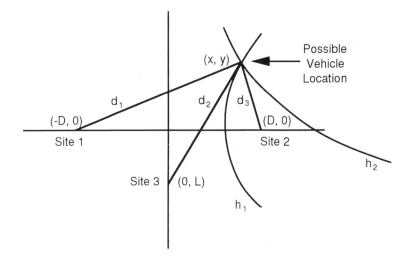

**Figure 9.4** Location determination by time distance of arrival (TDOA) or hyperbolic positioning.

As in the discussion of AOA, we assume that the radio-wave velocity is approximately equal to the speed of light (which may not be true in certain applications). Substituting the unknown coordinates of the vehicle and the known coordinates of sites 1 and 2 (as shown in Figure 9.4) into the previous equation, we obtain

$$\frac{x^2}{a^2} - \frac{y^2}{b^2} = 1$$

where $a = 0.5c\Delta t_{12}$ and $b = 0.5(4D^2 - c^2\Delta t_{12}^2)^{1/2}$. This is a hyperbolic function with the two sites as foci of the hyperbola. (The reader may recall that a hyperbola is a collection of points with a constant difference between the distances to each focus.) Similarly, we can derive another hyperbola $h_2$. The intersection of these two hyperbolas is the vehicle location.

Use of the TDOA technique for real-time location calculations requires fewer antennas and is less susceptible to signal blockage, or multipath reflection than using AOA. The main disadvantage of TDOA is the requirement for maintenance of a synchronized time source between all base sites. It may be difficult to implement and maintain the multisite synchronized time keeping accuracy required to measure the propagation of an RF signal. Note that this may not present a problem to the CDMA-based network at all since its sites have already been synchronized. Radio waves have a speed of approximately 300 m/$\mu$s (984 feet/$\mu$s), so that a 1-$\mu$s (one millionth of a second) time error in a single site could place a vehicle 300m away from its actual location. Most location systems require a position accuracy of less than 300m. One technique to solve this time-synchronization problem is to use GPS receivers as discussed in Section 3.4.2.

Another problem with TDOA is that channel bandwidth may impact the performance of this technology. The time difference measurement in TDOA may be affected by the narrow channel bandwidth since high-resolution time measurement requires a narrow pulse (or equivalent), and the narrower the pulse the greater the bandwidth required. By contrast, narrow channel bandwidth is not a problem for the AOA technology. This makes TDOA less accurate in narrowband analog systems than in wideband systems. To improve overall accuracy of location, some implementations have attempted to use a hybrid of the two technologies (TOA/AOA, AOA/TDOA, etc.).

Many terrestrial radio systems can be used to determine location based on the trilateration technology. As we learned, this technique relies on detection of a radio signal from three or more fixed reference points. Some of the better known radio navigation systems are DECCA, Omega, and LORAN-C, but they are much less popular than GPS for vehicle applications. Unlike LORAN-C, the first two systems use phase differences to determine locations.

The DECCA navigation system operates in the 30- to 300-kHz band [5]. It uses a series of three or four transmitters. These transmitters send out low-frequency phase-locked signals that allow a position to be determined from a location of the receiver in the resulting hyperbolic lattice. This position can be translated into a latitude-longitude coordinate or grid reference. The navigation system is mainly used in Europe and run by individual host governments.

Omega is a global navigation system used primarily by maritime and aeronautical users [6]. It operates at very low frequency (10.2 to 13.6 kHz) and was developed initially for use by the U.S. Navy submarine service. Control of the system has since been operationally extended to include a multigovernment partnership. This system has been used with differential techniques for vehicle location purposes in the United Kingdom.

LORAN-C is the "C" configuration of the Long Range Navigation (LORAN) system operated by the U.S. Coast Guard. The "A" configuration of this system was initially used for experiments prior to 1942. The LORAN-C system utilizes three or more land-based transmitting stations whose signals can be used by a receiver to perform hyperbolic position determination. The transmitters use frequencies between 90 and 110 kHz. By 1979, LORAN-C transmitter "triads" had been constructed to cover all North American coastal areas, Northern Europe, the Mediterranean, and the Asian/Pacific coast [7]. Today, LORAN-C provides services to most of the northern hemisphere [6]. For land-based applications, LORAN-C position errors may be as large as 500m. Differential LORAN-C has been developed in an effort to improve position accuracy. Meanwhile, other improvements have been made in LORAN-C during the last few years [8]. The different characteristics and complementary nature of LORAN-C and GPS suggest that it might be useful to integrate them for vehicle navigation [9]. Further information on DECCA, Omega, and LORAN-C and their applications can be found in [6,10].

On the whole, these terrestrial-radio-based systems are very efficient for maritime and airborne applications. This is partly due to the absence of large obstructions in these environments. On land, however, utilization of these systems results in reduced accuracy because of environmental and man-made obstacles such as mountains and large buildings. These types of obstacles cause signal attenuation and multipath reflection effects, which can limit the effectiveness of a land-based receiver for location determination. For instance, the location errors of LORAN-C and Omega may run as high as 500 and 2,000m, respectively. The ability of a receiving system alone to compensate for these signal distortions is very limited. On the other hand, terrestrial-radio-based systems allow voice and data communications to be merged into one system. Unlike satellite-based systems, some of these systems work inside buildings (as discussed below).

Another interesting form of terrestrial-radio-based location technology has been developed by Pinterra [11,12]. This system uses signals from commercial FM radio stations in conjunction with a reference station to calculate the location. This technique uses the pilot tones of FM stations (which are generally in the 19-kHz range) to calculate the location (see Figure 8.6 for the pilot tone signal). The location is determined via triangulation: The mobile receiver converts the phase measurements to the range measurements based on the signals received from at least three radio stations. This technology requires installation of a reference station (observer) at a known location in each metropolitan area; this station plays a role similar to that of the reference station in DGPS. The reference station calculates phase and frequency drift corrections for each FM radio station, these corrections are transmitted to the mobile receiver over an FM subcarrier or other broadcast medium to synchronize the transmissions and stabilize the frequencies. This technology has the advantage of wide coverage due to the high concentration of FM stations in many countries. FM stations can often cover up to 20,711 km$^2$ (8,000 square miles). Additionally, since FM broadcasts utilize frequencies (87 to 108 MHz) that are lower than GPS or cellular networks, the signal is less affected by obstacles such as buildings or hills. Because the FM signals can penetrate into buildings, this technology can be embedded with many portable device often used indoors. The system has a claimed accuracy of 10 to 20m.

Most terrestrial-radio-based systems generally have a minimum error of at least 150m. The worst systems may have errors of up to 2,000m. As we mentioned earlier, the performance of these systems often deteriorates when approaching urban canyon areas. More studies or hybrid methods (various combinations of the three approaches discussed or with other approaches) are needed to further improve these technologies. For instance, one method to reduce the positioning error would be to implement filtering techniques in the mobile device, which uses the relationship between carrier Doppler phase measurements and range estimate changes. Interested readers can find information on various other radio location technologies and systems in [13,14]. Many of these technologies have very promising potential for use in vehicle location applications.

### 9.2.3 Satellite Technologies

Due to the large investment in infrastructure, the number of satellite-based radio location systems is still limited. In this section we briefly discuss two such already used systems. The first system is the global positioning system (GPS). The second system is the EutelTracs operated by Eutelsat. These systems utilize line-of-sight signal reception and the TOA concept to calculate the position of the receiver. Information about other similar systems such as GLONASS and STARFIX can be found in [10,15,16]. Some of them as well as many GEO and LEO systems have been mentioned in Sections 3.4.2 and 8.3.8.

As we saw in Section 3.4.2, GPS is an all-weather, radio-based, satellite navigation system. It allows users to determine three-dimensional position, three-dimensional velocity, and time. The system consists of three segments. The space segment consists of a constellation of 24 satellites. The user segment consists of all the GPS receivers and support equipment. The ground segment is composed of the master control center and many widely separated monitoring stations. Ground control tracks the satellites, determines the orbits, and periodically sends correctional information and other data to the satellites for retransmission to the user segment.

A very simple location system can employ the GPS receiver alone as the position sensor. As discussed in Section 3.4.2, the accuracy of GPS receivers is degraded under selective availability (SA), and much better accuracy can be obtained by utilizing differential correction techniques (at an additional cost). Another approach is to use a dual GPS/GLONASS receiver, which takes advantage of 48 satellites in space (instead of only 24 for a receiver based on GPS satellites alone) and should provide position fixes most of the time, even in urban canyons. Note that GLONASS is not affected by SA. Detailed discussions of the underlying principles and methods for integration with other techniques can be found in Chapters 3 and 4. Note that as of this writing, the U.S. government has decided to conditionally phase out SA in 4 to 10 years, subject to annual review by the U.S. president. This will, of course, improve GPS accuracy for civilian users to a significant extent. However, note that SA may be reinstated at any time (national emergency, war, etc.) and this risk should be considered in the overall system design.

In contrast to GPS, in which the users only receive data (unidirectional transmission), the European satellite location service, EutelTracs, allows for two-way communications. The system uses an on-board terminal to facilitate this bidirectional capability. Vehicles communicate with two satellites, one used for master communications and the other for ranging calculations. The satellites used are in geostationary orbit 36,000 km above the Earth's surface. The reported accuracy of this system is on the order of several hundred meters [17].

### 9.2.4 Interface Technologies

The displays used in simple location systems can vary in capability. Design issues to consider are primarily cost, application, and operating environment. Simple text-based LED or LCD displays (as discussed in Chapter 7) are a low-cost means for providing low-bandwidth information to the vehicle operator. These types of displays are suitable for display of positioning information, traffic information, and limited route guidance information.

Other display subsystems that may be incorporated into a simple location application include a graphical map display showing the vehicle's position as a dot on the map. This type of display may or may not allow the user to enter information through a touch screen or other input device. More complicated systems require additional capabilities for user input (see Chapter 7 for additional details on this subject).

## 9.3 VEHICLE NAVIGATION

Vehicle navigation systems can consist of several integrated system building blocks as shown in Figures 6.1 and 9.5. The five shaded blocks in Figure 9.5 represent devices that can be implemented either as stand-alone components for navigation or as multifunction components that share some duties with entertainment-oriented equipment or other non-navigation equipment. The antennas used by the system can be implemented either separately or integrated into a smart device.

In general, systems created from complicated navigation building blocks tend to stem from requirements that specify features such as greater position accuracy, sophisticated human-machine interfaces, powerful guidance abilities, or integrated radio location receivers. As a result, these complicated navigation systems tend to be more expensive than their simple location counterparts.

### 9.3.1 Coping With Complex Requirements

In this section, we discuss the system issues associated with the implementation of a complicated navigation system. A typical navigation system may consist of a few or all of the modules depicted in Figure 6.1. Many commercial in-vehicle navigation systems are already available on the world market. These systems are generally implemented using the modules discussed in Part I. Interested readers may find literature on them at ITS-related trade shows and conferences, from magazines, or

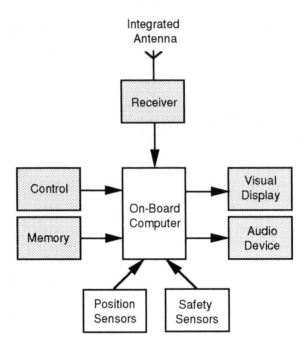

**Figure 9.5**  Complicated in-vehicle navigation system.

from [18–20]. Because we have already described the basic building modules in detail and will use a complex in-vehicle navigation system constructed from these modules as a case study in Section 11.3, we do not repeat that discussion here, but instead take a look at complicated navigational systems from a slightly different angle.

A thorough understanding of user requirements is a must for selecting and integrating proper navigation modules. The more complex the system requirements, the more complicated the resulting integrated system. For instance, there is a wide variety of choices, even for something as simple as a visual display unit, ranging from a simple black-and-white LED displaying only a few characters to a fancy active-matrix color LCD displaying text and graphics. Our discussion will be illustrated by walking through the system design issues that result from the functional requirements associated with a hypothetical product.

We consider the design of a rental vehicle utilizing an autonomous navigation system. In this design, we assume that the system incorporates a radio location receiver. Because this is an autonomous system, we further assume that the vehicle does not have any integrated wireless communications capability that can be addressed from a centralized facility. Without this communications link, it is obvious that all vehicle navigation functions must be performed by the system on board the vehicle.

Functional requirements vary from application to application. Some systems might require high position accuracy, whereas others will sacrifice accuracy for unit cost. Increased sophistication in one subsystem can make up for deficiencies in other subsystems. For example, as we saw in Chapter 4, a conventional or fuzzy logic map-matching algorithm can be used to overcome accumulative position errors by determining reasonable best fit locations on a map.

We begin this study by reviewing the following list of high-level functional requirements for the rental vehicle example:

1. The system shall have the capability to determine its current position within 20m of actual location over 90% of its travel time.
2. The system shall have the capability to translate its current position to a map coordinate, and subsequently to a best fit road segment location.
3. The system shall have the capability to display its current position on a map that can be viewed by the vehicle operator.
4. The system shall have the capability to accept a trip destination request and plan a best route to that destination.
5. The system shall have the capability to output audible and visual instructions corresponding to the directional maneuvers required to complete the planned route.
6. The system shall have the capability to determine when it is "off-route," that is, off the planned road segment.
7. The system shall have the capability to correct "off-route" conditions by generating a new route starting at the current erroneous position.

If we look back to our simple hand-held navigation device mentioned in the beginning of this chapter, we can see that it is not capable of satisfying any of these high-level system requirements. At this point we should analyze our requirements and consider their implementation, along with some of the resulting architectural and design issues. Let us begin with the first high-level requirement.

*The system shall have the capability to determine its current position within 20m of actual location over 90% of its travel time.*

We actually have two constraints on our system architecture as a result of this requirement. The first constraint implies a position solution within 20m of actual vehicle position. This constraint exists so that the system can differentiate between closely spaced road segments. An example of two closely spaced road segments might be an expressway and corresponding frontage road. The 20m resolution is required in order to minimize road-matching errors for closely spaced parallel road segments. The second constraint extends our first requirement by stipulating that we must be within this 20m limit during 90% of our operating time.

If we assume that a GPS receiver can operate at less than a 20m error (in the absence of SA) it may be possible to meet this 90% requirement. We would have to assume, however, that the satellite signals were not blocked for any more than 10% of our travel time. This is feasible if we are operating in a suburban or rural region without hilly or mountainous terrain. On the other hand, if we spend much of our travel time in an urban environment, it is very likely that we will not meet this requirement. Tall buildings, overpasses, and tunnels will block the GPS signal for significant periods of the travel time, which may in fact amount to more than 10% of operating time. This constraint implies that an additional positioning subsystem will have to be included to offset the deficiencies of GPS in urban environments. Because this additional subsystem should not be susceptible to the same weaknesses as the GPS subsystem, subsystems using similar radio location technologies will not be very useful. This means that a dead-reckoning subsystem is the most likely choice to augment GPS (assuming that a feasible mechanism is available or can be designed for the periodic correction of the dead-reckoning sensors by GPS).

Having two subsystems for positioning will of course increase the complexity of the system. Depending on the sensor fusion method used, some implementations will require arbitration to determine which positioning input takes precedence when the GPS subsystem and dead-reckoning subsystem do not agree. Map-matching methods can also be incorporated into the arbitration scheme to further improve the location accuracy since the next requirement already implies a requirement for a map database. Detailed discussions of sensor fusion and map matching algorithms can be found in Chapters 3 and 4.

*The system shall have the capability to translate its current position to a map coordinate, and subsequently to a best fit road segment location.*

Because we do not have a wireless communications link to a central facility, this requirement implies that the map database will need to be stored on an in-vehicle mass storage device. The database storage requirements for a reasonably sized metropolitan area (Section 2.6.1) suggest that we may need on the order of 10 to 200 MB of data storage (uncompressed), which includes additional room for possible future expansion of system coverage. Possible choices for storage devices might include CD-ROM drives, magnetic hard drives, or perhaps a PCMCIA memory card. Each of these storage technologies has advantages and disadvantages. The first two have the advantage of large storage capacity. The current capacities for these devices are well above probable storage requirements. Both devices share the significant disadvantage of utilizing moving parts. This disadvantage is due to the extreme operational environment intrinsic to vehicular applications. From extreme temperature ranges to frequent mechanical shocks, the environment is far from ideal for any storage device that utilizes moving mechanical parts, such as CD-ROM drives or hard disk drives. PCMCIA memory cards do not have this problem because there are no fast moving mechanical parts. However, the reduced storage capacity of the PCMCIA card relative to the other technologies is a disadvantage unless a technique can be developed for rapid compression and decompression. Because of the rapid advance of various mass storage devices, the drawbacks discussed here may not be a problem in few years, but it is still a challenge faced by today's design engineers. In addition to these problems, expensive storage and peripheral devices may very well increase the total system cost, and slow access devices may significantly affect the performance of the various embedded real-time software algorithms.

Assuming that we have selected a storage device, we can now have the system translate the reported position provided by fusion of the readouts from the sensors selected in the first requirement to the format used by the map database. This map database must be navigable, that is, provide information that can be used to plan a route and guide the vehicle. In addition, the map should have a position accuracy of at least 15m, so that it can be used to improve the inaccurate positions provided by the positioning sensors. The augmented GPS positioning subsystem may return vehicle coordinates (latitude and longitude). If the map database selected does not use WGS 84 as its ellipsoid reference as GPS does, we may need to transform the GPS-derived coordinates to the coordinates used by the map database or vice versa. Otherwise, displacements of up to 100m may occur. (Recall that Section 2.3 contains detailed discussions of map reference coordinate systems and transformations between various coordinate systems.) If the vehicle is traveling along a road, the system should be able to use a map-matching algorithm to determine the best-fit road segment. If there is no exact match, the algorithm should be able to find the nearest road segment and snap the vehicle onto it. Otherwise, it could simply use the sensor measurement to determine a position. Additional information on map-matching algorithms can be found in Chapter 4.

Regardless of the storage device selected, this system faces another significant drawback. The database used by this system is static in nature. We need to consider the problem that the database has not been dynamically updated to account for construction, road closures, new road development, and other modifications that might affect the database. This is particularly a problem for drivers of rental vehicles. The driver may be unfamiliar with some local roads, and is relying on the fact that the planned route will not be interrupted or delayed by road closure or construction detours. With a wireless communications capability, this problem can be addressed by incorporating dynamic traffic information. Without a wireless communications capability (as in our case), updates of road networks would have to be accomplished via a less dynamic mechanism, such as periodic map database releases. Even via this mechanism, there is no way for the newly released databases to have current road accident and emergency information. Unless we increase the project budget and add additional components to the system (as we will do in Chapter 10), we must warn the user of the deficiencies in the system. Of course, even the best dynamically updated database might contain some errors, so error recovery is always an issue in any design. If possible, we should make the system design as modular as possible to facilitate future expansion in later versions of the product.

*The system shall have the capability to display its current position on a map that can be viewed by the vehicle operator.*

Given this requirement, we can assume that the navigation system will need to include a graphical display, as described in Chapter 7. Again, we need to consider a significant list of human factors issues. Specifically, should the display be backlit? How will the display be affected by glare in daylight? Will it be bright enough to read in the direct sunlight? Where will the display be located? Will it block the driver's vision or cause the driver to divert attention from the road for significant periods of time? Will the display block either the driver's or front seat passenger's airbag safety device? All of these issues must be resolved before installation of a graphical display device.

As we have studied already, many display devices are available. They are based on different technologies and each has its own advantages and disadvantages. Because of the severe operational environment in the vehicle, impressive display units often have high price tags. A tough decision must be made in the trade-off among display quality, user convenience, and the price of the unit because this device may be one of the most expensive components in the whole navigation system.

*The system shall have the capability to accept a trip destination request, and plan a best route to that destination.*

This requirement is actually a multilevel requirement. The first point we need to consider is the issue of the method used by the operator to input the destination. Is this request done verbally? If not, can we use simulated keyboard entry via a set of switches or a touch screen to enter the street address, business name, or road intersection as a destination? If we allow users to enter addresses as their destinations, we need to make sure that the system can correlate street addresses with digital map coordinates if the database does not contain the coordinates of every possible address. This requires that the map database have attributes of street names and address ranges for each road segment. One method involves matching the desired address along the road using linear interpolation based on the address range for the road segment. The deficiency of this method is that address numbers are not always evenly distributed along a road segment, so very accurate address matching is not feasible.

Once we have settled on a means for inputting the destination information, we need to evaluate the second level in this requirement. This second stage implies that a sophisticated route-planning capability must be implemented. What options do we present the user in terms of how the route should be planned? Some considerations might be given to offering the user a choice among several planning (or route selection) criteria as follows:

- Fastest trip (time);
- Shortest distance (distance);
- Maximize or minimize expressway use (complexity).

Different planning criteria may require different attributes from the digital map database and evaluation functions defined in the algorithm. We also need to decide whether to allow users to select these criteria themselves or to stick with one criterion in the design without giving any flexibility to users. Note that since the system does not have dynamic traffic information, the fastest time choice listed by the system may not be accurate. A detailed discussion of various route planning algorithms can be found in Chapter 5.

*The system shall have the capability to output audible and visual instructions corresponding to the directional maneuvers required to complete the planned route.*

Like the issues involved in determining how to input the requested destination, we need to consider how the maneuvers associated with a planned route will be presented to the user, as described in Chapters 6 and 7. One issue that is encountered is whether we incorporate an audio subsystem to present instructions to the driver like "turn left at next traffic light," or "exit at ramp 256 east," etc. If so, does the audio subsystem have a volume control so it may be heard over traffic or simultaneous radio entertainment? Visually, how do we tell the vehicle driver to take a "hard" versus a "slight" turn? Such intersections might, for example, consist of both 45- and 135-degree right turns, or of a complicated roundabout.

An additional consideration concerns whether the maneuver is displayed on a separate visual direction screen (turn arrow) or as part of a road segment highlighted on a map (route map). Should the display stick with one consistent format, or use display screens that are sensitive to route guidance context? We may also want to use audible cues such as tones to indicate when to perform the maneuver, etc. All of these considerations will affect the appearance of the device and the acceptance of different user groups, because the display unit will impose the first and probably the most important impression on any user.

*The system shall have the capability to determine when it is "off-route," that is, off the planned road segment.*

During the route guidance process, we need to determine whether we are actually off-route, or just experiencing an error condition associated with the positioning subsystem. If the position report supplied by the navigation system does not indicate that the vehicle is on a road segment, a heuristic algorithm may be needed to determine whether the position measurement of the vehicle is in error, or the vehicle has left the road purposely to stop at an off-road location, such as a parking lot, driveway, etc., or whether there might be a database error. We expect the system to automatically detect the problem and perform appropriate action (such as warning or replanning), as suggested in Section 6.3. We may also need to decide whether a manual intervention or resetting mechanism is required to help the system recover from abnormal situations.

Given the static nature of our database, construction detours will probably be interpreted by the system as an off-route condition. A user-configurable mechanism may be needed to allow these off-route conditions to be ignored so that the system does not continuously warn the driver or try to replan a new route when the vehicle is on a detour due to road closures or construction.

Finally, we evaluate the last requirement of the example system.

*The system shall have the capability to correct "off-route" conditions by generating a new route starting at the current erroneous position.*

Assuming we have encountered a valid off-route condition, the system should recover and replan a new route based on the current position of the vehicle. How should the system warn the driver? Is the replan done in a timely fashion, or does it require an excessive amount of time? What maneuvering directions, if any, should be presented while replanning operation occurs?

We also need to decide how and in what format the system should present the results of a successful replanning operation to the user. In the case study presented in Chapter 11, the system uses the criterion that the new route must be at least 2 min shorter than the original route. Any route that does not satisfy this criterion will not be presented to the user. Other criteria have been used, such as 30% shorter than the estimated trip time for the original route [21]. We may want to conduct several experiments on our own in order to identify the best criterion for this particular project.

As can be seen, this example touched on a fairly large number of design issues. Even though we have not mentioned them, many other issues and low-level trade-offs must also be considered when designing a system of this type. These trade-offs can be inferred from the information contained in the previous chapters. Once we have identified the functions of the system and the proper hardware components to support them, we must take another look at the entire design to make sure that it fits the price range of the expected market and the proposed schedule. Several more iterations are generally required to finally settle on a design that all design team members can agree to. The design team should include key engineers, marketing personnel, and management personnel. Remember, it is generally worth it to spend more time on system specifications and requirements rather than less.

### 9.3.2 Dual-Use Navigation and Entertainment Components

Vehicle electronics evolved rapidly during the past decade, to the point where they have now been integrated into the vehicle architecture as integral parts of a sophisticated network. It is natural that many navigation components can be shared with entertainment equipment.

The shaded blocks in Figure 9.5 can support either navigation or entertainment or both. An integrated transceiver might be designed for use with AM/FM radio, GPS, cellular, and paging signals [22]. To reduce cluttered and inconvenient control devices, voice-activated controls, reconfigurable steering wheel controls, and reconfigurable feedback displays can be developed. We know that CD-ROM devices, hard disks, or memory cards can be used for external memory. For instance, a CD-ROM player could be used for storing the digital map database and navigation software as well as for playing music. In addition to serving as a memory device for the navigation system, a memory card reader could also serve other mobile office devices. The display monitor could be used for navigational maps or for commercial TV stations. Similarly, the speaker could be used to listen to turn-by-turn maneuver instructions, ordinary AM/FM broadcasting, or a hands-free cellular phone.

Many available products have combined both navigation and entertainment in a common platform. For instance, the Telepath 100 radio can direct users to a predetermined destination while still supporting all the functions of a normal radio [23]. In this system, the vehicle location is determined using GPS in combination with dead-reckoned positioning and an area map encoded on a PCMCIA memory card. The route is displayed to the driver using directional arrows. AudioNav uses the audio CD player to store navigation data and play music [24]. Many Japanese in-vehicle display units can be used for both digital maps and commercial TV [20].

Before closing this chapter, consider the following question. Recall from Chapter 6, a Braess paradox effect could occur if numerous vehicles equipped with identical navigation units receiving the same dynamic traffic updates were operating on a particular road network. What would happen if the usage of autonomous navigation-equipped vehicles is overwhelming in a popular tourist area? Would the effect be any different if these vehicles all had identical navigation systems or if they had different navigation systems?

## References

[1] P. H. Jesty and J. Giezen, "It's Architecture Jim, But Not as We Know It," *Traffic Technology International*, June/July 1996, pp. 26–30.

[2] Z. Liu, *Natural Beacon Guidance System, A New Kind of Land Vehicle Navigation System with Computer Vision*, Wuhan, China: Wuhan University of Technology, May 1990.

[3] M. Jenkin, E. Milios, P. Jasiobedzki, N. Bains, and K. Tran, "Global Navigation for ARK," *Proc. IEEE/RSJ International Conference on Intelligent Robotics and Systems*, July 1993, pp. 2165–2171.

[4] R. C. Johnson, "Neural Technology Put on the Map," *Electronic Engineering Times*, Apr. 15, 1996, pp. 1, 16.

[5] E. N. Singer, *Land Mobile Radio Systems*, Englewood Cliffs, NJ: Prentice Hall, 1994.

[6] P. Enge, E. Swanson, R. Mullin, K. Ganther, A. Bommarito, and R. Kelly, "Terrestrial Radio Navigation Technologies," *Navigation: J. Inst. Navigation*, Vol. 42, No. 1, Spring 1995, pp. 61–108.

[7] E. N. Skomal, *Automatic Vehicle Locating Systems*, New York: Van Nostrand Reinhold, 1981.

[8] G. L. Roth, T. Blandino, and P. Schick, "New LORAN Receiver Technology Significantly Improves Overall System Performance and Substantiates LORAN Viability as GPS Backup," *Proc. ION National Technical Meeting*, Jan. 1996, pp. 219–234.

[9] G. Lachapelle and B. Townsend, "GPS/LORAN-C: An Effective System Mix for Vehicular Navigation in Mountainous Areas," *Navigation: J. Inst. Navigation*, Vol. 40, No. 1, Spring 1993, pp. 19–34.

[10] B. Forssell, *Radionavigation Systems*, New York: Prentice Hall, 1991.

[11] J. Cisneros, D. Delley, and L. A. Greenbaum, "An Urban Positioning Approach Applying Differential Methods to Commercial FM Radio Emissions for Ground Mobile Use," *Proc. ION 50th Annual Meeting*, June 1994, pp. 83–92.

[12] D. C. Kelley, D. T. Rackley, and V. P. Berglund, "Navigation and Positioning System and Method Using Uncoordinated Beacon Signals," *United States Patent No. 5173710*, Dec. 1992.

[13] C. J. Driscoll, *Survey of Location Technologies to Support Mobile 9-1-1*, C. J. Driscoll & Associates. Also published by State of California Department of General Services, July 1994.

[14] P. J. Duffett-Smith, "High-Precision CURSOR and Digital CURSOR: The Real Alternatives to GPS," *Proc. Third International Conference of the Royal Institute of Navigation and the German Institute of Navigation (EURNAV '96)*, Paper No. 17, June 1996.

[15] W. F. Blanchard, "Civil Satellite Navigation and Location Systems," *J. Navigation*, Royal Institute of Navigation, Vol. 42, No. 2, May 1989, pp. 202–220.

[16] E. Kaplan (Ed.), *Understanding GPS: Principles and Application*, Norwood, MA: Artech House, 1996.

[17] T. Scorer, "An Overview of AVL Technologies and Application," *IEE Colloquium on Vehicle Location and Fleet Management Systems*, June 1993.

[18] R. Cogan, "Wheels in Motion," *ITS World*, July/Aug. 1996, pp. 18–22.

[19] R. L. French, "Navigation Aids and Intelligent Vehicle-Highway Systems," in *Automotive Electronics Handbook*, R. Jurgen (Ed.), New York: McGraw-Hill, 1995, Chap. 29.

[20] E. J. Krakiwsky and R. L. French, "Japan in the Driver's Seat," *GPS World*, Oct. 1995, pp. 53–60.

[21] C. Blumentritt, K. Balke, E. Seymour, and R. Sanchez, "TravTek System Architecture Evaluation," *Federal Highway Administration*, Report No. FHWA-RD-94-141, July 1995.

[22] W. C. Spelman, "Entertainment and Information Merge in the Automobile," *IEEE Vehicular Technol. Soc. News*, Feb. 1996, pp. 49–55.

[23] E. Y. Wu and D. L. Welk, "An Alternative Approach to Automobile Navigation," *Proc. IVHS America 1994 Annual Meeting*, Apr. 1994, pp. 50–54.

[24] J. Smolders, R. Diller, and D. Van Compernolle, "Noise Robust Speech Recognition Makes In-Car Navigation Safe and Affordable," *Proc. Second World Congress on Intelligent Transport Systems*, Nov. 1995, pp. 601–604.

# CHAPTER 10
▼▼▼

# CENTRALIZED LOCATION AND NAVIGATION

## 10.1 INTRODUCTION

A centralized location and navigation system utilizes some combination, or subset, of the building blocks discussed in Part I. These building blocks can be integrated into both an infrastructure and the vehicle capable of performing dynamic multivehicle location and navigation processing. In this chapter, we attempt to discuss the issues related to various configurations of these systems. From a system-architecture standpoint, three high-level functional blocks are required to perform centralized location and navigation, as shown in Figure 10.1.

The host consists of one or more facilities with the capability to determine the location of or to provide guidance/advisory information to one or more mobile vehicles or devices. For a simplified, simplex (one-way mobile to host) location-determination capability, this component might include a location function, a human-machine module, and an interface to a wireless communication module. In a more complex arrangement, the host may provide full-duplex (two-way between mobile and host) navigational support, complete with traffic data fusion to provide dynamic route guidance based on real-time traffic information. The communications network provides the conduit between the host and the mobile unit. This component has the capability to provide either active or passive support to the location function. As

---

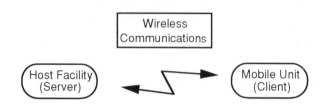

**Figure 10.1** Architectural blocks for a centralized location and navigation system.

described in Chapter 8, numerous communications network configurations supporting various throughput levels can be incorporated into this architectural component.

The mobile unit may also function at various levels of location and navigation sophistication. In a simple configuration, the signal radiating from the antenna used by a mobile cellular phone can be used to determine the location of the mobile unit via triangulation data interpreted by a central facility using multiple antennas (as discussed in Section 9.2.2). The operation of this location determination system may be transparent to the mobile handset user. In this simple case, the primary responsibility for determination of mobile unit location rests with the host. In a more complex scenario, the mobile unit will take advantage of one or more positioning capabilities, such as a GPS receiver combined with dead reckoning and map matching. The position derived from the combination of these positioning capabilities is transmitted through the communications network back to the host. The host may then communicate back traffic or dynamic route guidance information directed specifically to the mobile unit.

The applications supported by centralized location and navigation systems are several fold. One important application that can be supported by these systems is a wireless enhanced 911 (E911) location capability. *Wireless E911* is a system with the ability to locate a caller who uses a wireless phone to request emergency assistance (Section 10.4). Ideally, this application would provide the same functionality as landline-based implementations. Widespread implementation of this capability would probably rely on a location technology that is primarily supported by centralized host facilities and communications infrastructure. The extent of network infrastructure modifications needed to support this capability, weighed against the accuracy of the location determination, becomes an important design trade-off in terms of cost and duration of customer base roll-out. A nonexhaustive list of additional applications for centralized location and navigation technology is provided:

- Emergency private wireless E911;
- Emergency vehicle dispatch and tracking:
  - Police;
  - Fire;
  - Ambulance.

- Public transportation:
  - Bus fleet.
- Private fleet management:
  - Trucking (freight and asset tracking);
  - Taxi services;
  - Trains.
- Secure delivery fleet:
  - Armored vehicles and couriers.
- Private vehicle services:
  - Emergency and roadside assistance;
  - Travel information (route guidance, traffic, weather, gas, food, lodging, etc.);
  - Theft recovery.

The variety of potential applications given in the preceding list for centralized location and navigation systems indicates the need for a diverse set of business and engineering solutions. These solutions tend to have varying economic and infrastructure constraints specific to each particular application. Three main system design issues are associated with these constraints:

- Placement of the location and navigation capabilities;
- Accuracy of locations and frequency of location updates;
- Selection of wireless communications technologies.

The primary issue we need to consider is that of where the place will be for residing the location and navigation capabilities. For example, consider a mobile cellular system manufacturer intending to implement a wireless E911 feature in cooperation with a service provider. This feature is designed to provide the mobile caller's location to the dispatcher (or operator) at an emergency center. The manufacturer may consider a redesign and retrofit the mobile handsets to incorporate the intelligence required for the location determination. The redesign could, for example, involve adding a GPS receiver to the handsets. The cooperating service provider must then consider the time and cost associated with a recall of the existing customer base. On the other hand, the systems manufacturer may decide to retrofit its base cell sites to incorporate TOA, AOA and/or TDOA location technologies as discussed in Section 9.2.2. This places the intelligence of the location determination at the base cell sites, instead of in the handsets. A design using this cell-site retrofit approach may require the addition of "smart" antennas in addition to some level of effort and cost associated with reprogramming the sites. The same considerations hold for navigation, that is, do we design a system with a centralized or distributed navigation ability? In the distributed-approach case, the navigation capability must be resided on the mobile unit.

As far as location accuracy is concerned, we saw in the previous chapters that a wide variety of different location technologies are available. Each has its own accuracy range, and we must select among them based on the system requirements and applications. Table 10.1 lists typical performances for these technologies. This table attempts to give readers a rough idea of the potential of each technology. In actual usage, the performance may depend on a variety of other factors. In fact, many of these technologies will improve over time, and therefore their characteristics will change too. The location technology selected and implemented must satisfy system requirements (as discussed in Section 9.3.1). Similarly, for applications depending on remote location updates, the wireless data update rate needs to be carefully chosen in order to achieve the required system specifications. As we saw in Chapter 8, the specific wireless communications technology selected (and its attributes) has a significant impact on the cost, reliability, and accuracy of the overall system.

In the following sections, we discuss the design and implementation issues associated with centralized location and navigation systems. For convenience of discussion, we first study location technologies and then discuss how navigation technologies can be added to the system. In each category, the centralized approach (placing intelligence on the host) leads the discussion before the distributed approach (placing intelligence on the mobile unit). Despite this distinction, readers should keep in mind the fact that some technologies and systems may not fit into this neat classification. Hybrid approaches on variations of these approaches do exist. All of these approaches and systems contribute to the centralized location and navigation technologies we have today.

## 10.2  AUTOMATIC VEHICLE LOCATION

An automatic vehicle location (AVL) system tracks the positions of a fleet of vehicles in a particular area and reports the information to a host via a communications

**Table 10.1**
Performance Comparison of Various Positioning Technologies

| Technology | Performance |
|---|---|
| Dead reckoning (DR) | Poor (poor longer term, but good short term) |
| Terrestrial radio | Fine (150–2000m, will improve) |
| GPS | Fair (100–300m with SA and without blockage) |
| DR + map matching (MM) | Fair (20–50m without loss) |
| DGPS | Good (10–20m without blockage) |
| GPS + DR + MM | Better (15–50m continuous) |
| DGPS + DR + MM | Best (10–15m continuous) |

infrastructure. This host can take different forms, such as a dispatch center, a traffic information center, or a transportation management center. All of the technologies introduced in Section 9.2 may be used for AVL. The only addition to the technologies discussed there is that vehicles used in AVL systems must have a wireless communications component (module) to connect with the centralized host facility to which it periodically reports its position. An appropriate infrastructure must also exist to support the wireless communications needs of the system.

Determination of the location of each vehicle typically involves a RF transmission initiated by the mobile vehicle. In one type of system, the sole purpose of this communications system may be to transmit the actual location of the vehicle. In this case, the mobile unit plays an active role in location determination. For example, the transmission may contain a "position report" data packet, which includes the vehicle's latitude and longitude derived using a GPS receiver installed in the vehicle. On the other hand, the mobile unit may play a passive role in location determination. In this type of system, the actual information on the vehicle position may be derived as a by-product of a wireless E911 or a device in the vehicle that can be addressed by an infrastructure. We explore this passive approach to the location determination in the following section, and then consider utilization of the mobile unit as an active participant in location determination in a subsequent section.

### 10.2.1 Centralized Approach

A centralized location system is a system that uses a host computing facility and communications infrastructure to remotely locate or track a fleet of vehicles within a specific area. In short, the system uses a fixed host facility and infrastructure to determine the vehicle location. The host may take the form of an urban traffic information center or possibly an emergency vehicle dispatch center. Such a system generally includes the building blocks shown in Figure 10.2. Depending on the design, a simple AVL system may not need a sophisticated map-matching module at all. In addition, at least a simple communications component and a simple human-machine interface are required at the mobile end. Note that although our discussions are centered around vehicles, the fleet of vehicles in this system could be a group of hand-held mobile devices.

The host in a centralized location system will generally provide the capability to monitor and possibly dispatch all members of a specific fleet of vehicles. Each of the vehicles in the fleet will have the capability to periodically signal the host via an RF transmission. As a member vehicle of the specified fleet moves within the area monitored and controlled by the host, it periodically transmits or receives a wireless communications signal, which is then processed by the communications network and subsequently delivered to the host facility.

This centralized location capability often requires using the communications network as an active component of the location determination process. For example,

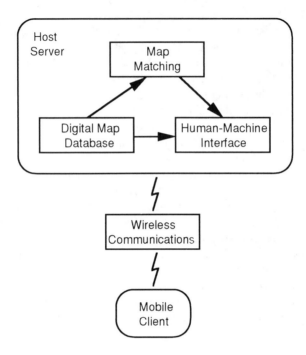

**Figure 10.2** Generic centralized AVL system.

since we have a large installed base of vehicles equipped with cellular handsets, and a mandatory requirement to implement location capability in a wireless E911 system (Section 10.4), we should want to avoid retrofitting or replacing all handsets. Instead, we would look for a solution based on performing the location function at each cell site. Currently, three main techniques are used for performing infrastructure-based location determination (as we saw in Section 9.2.2). Recall that the TOA, AOA, and/or TDOA technologies can be implemented on either the host system or the mobile unit. In a centralized location system, all location capabilities are accomplished in the host infrastructure.

Assuming that we have calculated the position of the vehicle within a cellular site, the next step is to allow the host to process the vehicle position data and output the location of the vehicle on a map. For a simple AVL system, it is straightforward to place a dot representing the vehicle position on a map provided the map has been geodetically referenced. Either raster-encoded or vector-encoded maps can be used (Chapter 2). The user must live with the fact that the dot or icon may not be precisely on the road. For more accurate locating, a map-matching module can be used in conjunction with a vector-encoded digital map database. To fully exploit the map database already resident in the host, the system can use the known map information to filter out the errors associated with the calculated position. The system then translates the calculated vehicle position to map coordinates within the bounds of the monitored area. This translation is achieved via coordinate information available from the map database system.

Given a sophisticated map database system, the data for the reported vehicle position can then be associated with a location on a map display. However, as we discussed, the raw position data still need to be translated into points on the map for viewing. This function could be provided by a modified map-matching module (instead of the in-vehicle map-matching module described in Chapter 4) or other enhanced modules. When defining this mapped location, the errors related to uncertainty in the vehicle position need to be evaluated, and the best available estimate for the location of the vehicle should be derived using a statistical algorithm.

The host will generally utilize an extensive network of high-capacity computing systems. Given this capability, additional dispatching and traffic management capabilities may be incorporated into the host. Besides road segments, shape points, connectivity, road name aliases, street address mapping, and hierarchical level information, the host database can incorporate various other components as follows:

- Vehicle-specific information;
- Vehicle year-make-model;
- Vehicle identification number (VIN);
- Vehicle registration—owner information;
- Vehicle route plans (e.g., fleets for public transportation and delivery services);
- Last reported location (position and direction);
- Last position report time;
- Location-specific information;
- Business names and address (possibly including phone numbers and hours of operation);
- Residential names and addresses;
- Points of interest (POIs).

These and other additional attributes could be used by the host to communicate with the mobile unit either verbally (using operators) or automatically (using data channels). Typical services might include providing route guidance instructions and service information to any vehicle in the monitored fleet. This assumes that the mobile and communications components of the system are capable of handling these data services.

A large-scale centralized host has the capability to store or maintain vehicle position and direction information for an entire fleet of vehicles in order to assist in performing the function of plotting vehicles on a display screen. For example, if a series of position reports shows a vehicle to be operating at high speed on an interstate highway, and the latest position reported for the vehicle places it on a frontage road, the host responsible for tracking the vehicle should be able to validate whether any off-ramps to the frontage road were encountered between the last reported position of the vehicle and its current reported position. If no off-ramps were available, or if the speed limit of the frontage road was significantly lower than that of the interstate highway, the system could assume, with relatively high confidence, that the vehicle's actual position was still on the interstate highway and a position reporting error had occurred. These functions require a navigable digital map database with enough road-related attributes, as defined in Chapter 2.

Other "sanity checks" can also be incorporated into a host to account for system-wide architectural weaknesses. A good example of this would be a host using AOA technology for vehicle positioning. Recall that RF signal blockage and/or multipath reflection may place a vehicle in the opposite direction from the actual vehicle position (Figure 9.3). If this scattering problem is localized to a small geographic area, the map-matching or map database module on the host could use RF profile information to account for and correct this signal path anomaly. The complexity and accuracy of the module and its ability to translate a vehicle's position to a map location is dependent on the resources of the host, the size of the vehicle fleet, the geographic area covered, and the quality of the map itself.

As discussed, each vehicle position report is passed to the map database, map-matching, or human-machine interface module (depending on the system design); the map database information is then used to translate the position into a map coordinate. Once this translation has been accomplished, an icon representing the vehicle and its current location can be output to a host map-display station. The display of all vehicle position information is controlled by an augmented human-machine interface (instead of the in-vehicle human-machine interface module as before). This display can take many forms. For example, vehicles and locations may be displayed as a tabular list of street addresses or as an icon on a road map. A road map may display vehicle information in a form similar to that used by air-traffic-control stations to display aircraft (which combines text with a spatial representation of the aircraft position). The human-machine interface can also take on a wide variety of formats, as discussed in Chapter 7.

The display function of a centralized location system differs from that of an in-vehicle map display. In-vehicle displays are generally only concerned with the location of a single vehicle on a map. The host display function is concerned with displaying information on an entire fleet of vehicles, possibly simultaneously. This function may use various magnification or "zoom" levels to display local and detailed, or wider and less detailed, geographic areas. This allows multiple vehicles to be tracked, but it may be necessary to sacrifice the details of individual neighborhood roads in order to track a large number of vehicles in the area being covered. This zoom capability should be configurable under host operator control, and also automated if necessary in case vehicles move out of the edge of the coverage area. The host should also have the ability to track, in detail, the route of a single vehicle of interest. This ability is especially important when covering specific emergency vehicles (such as ambulances) responding to an incident.

In addition to TOA, AOA, and TDOA, other technologies can also be used for remote location determination by the host. For instance, short-range beacons or magnetic field loop detector sensors (Section 8.3.7) installed along fixed bus routes and controlled by a host facility may be sufficient to update location information for a public transportation management center. Some literature refers to this technology (and beacon technologies) as signposts and systems constructed from these technologies for automatic vehicle location as proximity AVL systems. Interested readers can find in-depth discussions and analysis of some proximity technologies in [1]. Another available technology is the TIDGET system [2,3]. Unlike traditional GPS-based location systems, the in-vehicle TIDGET receiver does not compute latitude and longitude from the satellite signals. Instead, the on-board receiver passes the raw satellite measurements through the communications network to a host for processing. The host then calculates the location of the mobile, assisted by a map and a reference GPS receiver. Because no location calculations are performed on the mobile unit, the in-vehicle equipment cost is reduced. For a survey of available AVL technologies and applications, refer to [4] and the other references mentioned in Section 9.2.2.

Up to this point we have primarily been concerned with vehicle location from the perspective of the host. The configurations discussed have assumed the intelligence of the location determination would reside in the host and communications network. This approach makes sense when we wish to avoid large-scale or high-cost retrofit/installation of equipment on an existing fleet of vehicles. Next, we will look at vehicle location systems where the intelligence of the location determination function resides on the vehicle itself.

## 10.2.2 Distributed Approach

A distributed location system is a system that can track the position of a fleet of vehicles within a particular area based on devices at the mobile end that report the information to a host at a central facility. In short, in this system, the individual vehicle or mobile device determines its own location. As before, the central host can take different forms but without the location determination capability. All of the location technologies introduced in Section 9.2 are applicable to this system.

The two most important modules for an AVL system are positioning and communications. As we now know, many different technologies are available for use in these modules. The positioning module could be as simple as a GPS receiver alone or as complex as a multiple-sensor (including DGPS), map-matching-based subsystem [5]. The communications module could be as simple as a one-way data-only link from the mobile unit to the host or as complex as a two-way data and voice connection. Furthermore, there are various communications media available for wireless connection of an AVL system (as discussed in Chapter 8): simple paging networks, cellular networks, land-mobile radio networks, short-range beacon networks, and satellite networks. The list of different combinations of technologies can go on and on. Figure 10.3 shows one representative class of generic AVL systems, where the human-machine interface could be used for a variety of purposes. One simple example is to display information for the driver. The simplest system could consist of only a positioning module for the mobile client. Note that the host server might have the exact modules shown in Figure 10.2 or just a map database with a human-machine interface.

The system architecture for this generic AVL system is somewhat similar to that described in the previous section, but with the position determination capability hosted on the mobile unit. It is assumed that the host can use the previously described components in addition to providing a server that responds with location information whenever a client vehicle provides position data.

Now we use a very simple AVL system as an example to see how such a system might work. In this system, the only location determination component is an on-board GPS receiver. To reduce the overall system cost, none of the map database and map-matching modules is supported in the vehicle. In this architecture, the

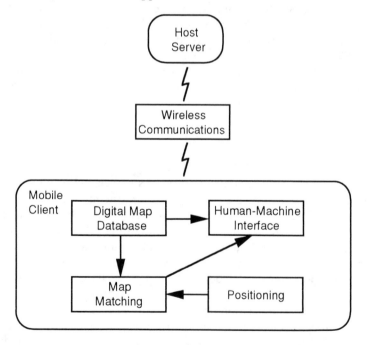

**Figure 10.3** A generic distributed AVL system.

mobile utilizes a simple 40-character LCD to display information and dispatch messages. Additionally, the mobile unit will query its position from the GPS receiver every 5 min and report this position to the host while receiving messages from the host. After obtaining the location report, the host will display this location information on its monitor, provided the vehicle in question is monitored. We assume that this system supports a maximum of 100 package delivery vehicles in a single fleet over an RF communications infrastructure.

First, consider the GPS-based positioning. Recall from Section 3.4.2 that the accuracy of the GPS receiver is affected by SA, so that uncorrected GPS measurements are not suitable for applications that require highly accurate position, velocity, and altitude information. However, in our case study we can assume that the package delivery vehicle is following a known route, and the position report it provides to the host only needs to be within a block or two. After receiving the reported position from the mobile GPS unit, the known route plus the position information will provide enough information for the host to obtain a rough estimate of the location (in this application, a rough estimate may be all that is required).

Once the vehicle being tracked has position data available from the GPS receiver, the position is reported to the host. This mechanism is provided by the mobile-based client. This client can be implemented as a software library capable of supporting multiple communications protocols and hardware architectures. The basic logic for this client is to initialize its communications channels and then enter a polling loop that periodically reads the communications channels. Note that we assume that there are two communications devices on board. One is a transceiver for communicating with the host via the wireless communications module and the other is a GPS receiver that provides position information for the on-board positioning module.

The channel initialization algorithm opens the inbound and outbound communications channels for the transceiver separately, implying read-only and write-only capabilities. A third separate initialization procedure is performed on the GPS receiver. The channel to the GPS receiver is read-and-write capable. This allows initialization commands to be sent to the GPS receiver, and responses and position data to be received from the receiver. It is typical for the initialization algorithm to have an automatic error recovery component in an embedded mobile system. This component may use staged recovery logic in an attempt to keep the system operating after a failure has occurred.

Next, we look at the mobile polling algorithm. Once the communications initialization has succeeded, the system will enter a continuous polling loop. A simplified algorithm for this polling activity might look like the following algorithm.

*Mobile Polling Algorithm*

1. Get the Current_Time.
2. Set the Next_GPS_Position_Query_Time = Current_Time + 5 minutes.

3. Until system termination, do the following:
   3a. Get the Current_Time.
   3b. If the Current_Time = Next_GPS_Position_Query_Time, read the GPS_Position and set the Next_GPS_Position_Query_Time = Current_Time + 5 minutes.
   - Format the GPS_Position_Message and send it to the host via the RF outbound channel.
   3c. Check the RF inbound channel for an inbound message and its type.
   - If the Inbound_ Message_Type is an Information_Message, format it on the LCD_Display.
   - If the Inbound_Message_Type is a Dispatch_Message, sound a tone and format it on the LCD_Display.
   3d. Go to step 3a.

In this polling algorithm, we assume that the mobile system has the capability to provide the current time obtained from an on-board source. We could, if necessary, derive it from the GPS receiver itself. However, if we derive the current time from the GPS receiver, our algorithm would have to change so that the GPS receiver is polled each time through the loop to obtain the current time. Regardless of how we obtain the current time, we would only report position data to the host every 5 min, in order to reduce the amount of traffic between the mobile unit and the host. This update rate could be set to some other fixed interval based on the application. At each of the reporting steps in step 3b, the algorithm creates a position message for the host in an agreed-on format, and sends it to the host via the outbound communications channel. At a minimum, this message will contain the vehicle identification, the latitude, and the longitude of the vehicle. The interval timer for controlling of the polling and reporting period is reset whenever the mobile unit receives a position report from the GPS receiver.

Another simplifying assumption made by this algorithm is that the GPS position report can be derived from a single reading every 5 min. To provide a reasonable position report, the GPS receiver must track at least three satellites to obtain a two-dimensional (latitude and longitude) fix. In an urban environment with tall building blockage, this condition may not be met at all times. Several seconds may elapse before the receiver is able to compute a position fix. The receiver may simply output the last known position if it has been on continuously.

Next, the algorithm checks to see if there is an inbound message from the host. This inbound message may be either an information message, a dispatch message, or some other type of message. For instance, information messages might contain information like "On Main St.—5 min behind schedule" or something of that nature. A dispatch message may be preceded by a tone to differentiate the message from an ordinary information message. The algorithm is also simplified in that there is no provision for the driver to acknowledge the receipt of a message, or prevent a new message from overwriting a potentially unread message. These issues would have to be addressed during system design.

In this case study we assumed that a wireless communications module was available to support messaging between the mobile unit and the host. The communications requirement for this small fleet of delivery vehicles, reporting at 5-min intervals and receiving information and dispatch messages, implies a fairly light-duty infrastructure. For example, the worst-case inbound message rate would be:

100 vehicles × 12 messages per vehicle per hour × 30 characters per message
= 1 message/3 sec
= 10 characters/sec
≤ 100 bps.

The outbound message rate would be different from the inbound rate, depending on the structure of the information or dispatch messages.

The type of communications infrastructure to be used depends on several additional factors. Some of these factors are coverage, equipment cost, installation cost, maintenance cost, and usage cost. Based on the size of the delivery fleet in the current case study, it is likely that all communications would be run by subscription through a communications service provider. On the other hand, a dedicated network is typically used for controlling a much larger fleets of vehicles or public emergency fleets.

One approach to supporting the communications infrastructure involves using cellular modems. These modems provide data messaging over the cellular network. This tends to be a higher cost method of message delivery in that today's cellular networks are oriented toward voice delivery. For analog circuit-switched data transfer, each inbound and outbound message would require a call to be established, a short message delivered, and, finally, the call disconnected. We could consider leaving the cellular connection up continuously, but this would be expensive and carrier loss due to changing RF conditions (fading) would be likely to occur in this mobile environment. In addition, cellular data modem selections depend on various factors, such as circuit-switched data transfer or packet-switched data transfer, and transmission data rates and protocols. Refer to Section 8.4 for a brief discussion, as well as [6] and the references contained therein for more information on modem selection in cellular-based AVL applications.

Other communications services such as those provided by several private data radio systems or specialized mobile radio (SMR) operators, tend to be oriented toward data delivery. The subscriber equipment costs tend to be significant, but pricing can be set up to be directly related to message throughput, which is low in this case study. An extended two-way paging system may be a reasonable communications solution for this type of application. Messaging rates, subscriber costs, and message delivery costs are a good fit with the communications requirements. In addition, the communications bandwidth limitation for the host, in this case, may be a function of the number of vehicles in a fleet attempting to transmit simultaneously to the host. We could reduce the communications cost and probably the communications bandwidth by reporting the mobile location back to the host only when necessary. One strategy is to use the communications channel based on the distance traveled rather than the time interval. This will eliminate wasted communications expenses in situations such as the vehicle is standing still and continues reporting its position back to the host.

Another alternative for the communications infrastructure may involve a hybrid combination of technologies. We may be able to use FM subcarrier broadcast for messaging from the host to the vehicles, and provide vehicle to host data through a one-way paging mechanism using packet switching.

The host (server) side of the communications capability would be implemented in a similar fashion to the mobile (client) side. The lower levels of the communications system would most likely queue incoming message reports, which would be processed in a first-in first-out manner. Each received message would be evaluated for vehicle identification and position (latitude and longitude). The received message data might have the following format:

NNN | XXX.XXXXX | YYY.YYYYY<CR>

where

| | |
|---|---|
| NNN | vehicle identifier number (VIN) |
| XXX.XXXXX | ASCII-numeric representation of the latitude |
| YYY.YYYYY | ASCII-numeric representation of the longitude |
| <CR> | message delimiter |

This message would be time-stamped by the host and passed to the human-machine interface, map database, or map-matching module, which could then use a database to correlate the vehicle ID number with a known vehicle route. This route information, combined with the reported position and delivery time, could be used to assist in accurately determining the vehicle's location along the route. If the latitude and longitude reported are not on the route, the module can then determine the closest route location and map the vehicle to that location. The time-stamp of the received message can also be used as a cue by correlating the vehicle to a known delivery schedule and destination. An outbound message indicating the delivery status of the vehicle can then be sent to the vehicle in response to the reported position message. This outbound message might have the following format:

NNN | T | M[40]<CR>

where

NNN     vehicle identifier number (VIN)
T     message type (information or dispatch)
M[40]     ASCII free form 40 character maximum text message
<CR>     message delimiter

This message would be sent to the mobile unit and displayed on a 40-character LCD. An example message might be as follows: 001 | "I" | "On Main St.—5 min behind schedule." The vehicle driver would only see the text message "On Main St.—5 min behind schedule" on the LCD display. Dispatch messages to the vehicles would take the same message format. It is assumed that we are using an FM subcarrier or packet data service that supports broadcast messages to our vehicle fleet. We could send a message to a single vehicle by using its identifier, an "I" for an informational message or a "D" for a dispatch message such as "Special Pickup at 365 Commercial Ave." A special vehicle identifier such as 000 or 999 could be used to indicate that the message is a broadcast message to be received by all fleet vehicles.

The host/dispatcher station might include a map with the vehicle locations displayed, as discussed in the last section. As discussed in Chapter 2, a preprocessing step may be necessary to translate the vehicle position data format into the format used by the map database. For example, we may need to take WGS 84 position information and translate it to UTM if the map database uses UTM encoding. This translation may or may not be necessary, depending on the system components used. The map database itself can be much more extensive if it resides on the host rather than the vehicle. In addition, the human-machine interface might provide an editing window for the dispatcher to enter vehicle IDs and messages in order to provide informative messages to one or more vehicles in the fleet.

In summary, we have outlined a case study where the intelligence of the location determination is embedded in the mobile unit, which implies a higher cost per mobile unit. We can, of course, increase system accuracy (at added cost) by including additional modules in both the mobile unit and host. For example, in the case studied here, adding more on-board sensors or map-related functions would allow the vehicle to report its position back to the host even when a GPS blockage occurs; this is a very valuable feature for mission-critical applications such as police vehicles [5]. Furthermore, with various modifications, an AVL system could provide location-dependent services, such as mayday, roadside assistance, stolen vehicle recovery, and even route guidance. In the last section, we discuss some applications of these traveler-oriented services.

## 10.3 DYNAMIC NAVIGATION

A dynamic navigation system uses real-time traffic information to assist users traveling on road networks. In the literature, this technique is also known as dynamic route guidance. The on-board hardware generally includes some or all of the components shown in Figure 9.5 with the exception that a transceiver is substituted for the receiver. As discussed in Section 6.4, it is much better to use dynamic travel costs than static travel costs for route computation, since the dynamic cost information reflects the actual situation on the roads. These real-time traffic data add one more dimension for the system to solve the route guidance problem, that is, time. In static navigation systems, the cost of each link (or road segment) is fixed. In dynamic navigation systems, the link cost varies with time as road and traffic conditions change. We shall first discuss systems in which the route guidance intelligence resides on the centralized host facility.

## 10.3.1 Centralized Approach

Centralized dynamic navigation systems rely on a multivehicle (system-wide) route-planning module to guide vehicles on the road based on real-time traffic information. In other words, optimizing each individual route for all the vehicles in a particular road network is solely the responsibility of the host.

Centralized dynamic navigation systems have a central road map database, which can be updated in real time by vehicles on the road acting as probes reporting back traffic information. This database can also be updated using other techniques and methods of analysis. These means include loop detectors, video images, ultrasonic devices, anecdotal information such as police reports and mobile phone user reports, etc. The system includes a centralized multivehicle route-planning module. As we saw in Section 6.4, the main difference between static route planning and dynamic route planning is that the former uses a static database for planning and the latter uses a dynamic database integrated with updated real-time traffic information. The central map database is updated periodically based on the current traffic conditions. The multivehicle route-planning module calculates optimal or near-optimal routes for every possible (or selected) origin-and-destination (O-D) pair or designated-zone pair in the road network. There are essentially two methods: on-demand calculation and periodic calculation. The former is calculated for each driver on request based on the O-D pair submitted by the driver. The latter is calculated periodically for all (or selected) O-D pairs or for designated zone pairs. Detailed discussions of various multivehicle route planning algorithms can be found in the references listed in Chapter 5. Note that multivehicle route planning was briefly introduced in Section 5.1, and a few algorithms were discussed in Section 5.6. After planning, these optimal or near-optimal route data are then transmitted to all vehicles or to designated vehicles at periodic intervals to direct them to their respective destinations. On the mobile side, a destination must be selected by the driver using an in-vehicle navigation device. This information can be used for planning, or (in the case where broadcast route data are being received) it can be used by the on-board system to filter out guidance data irrelevant to its destination.

Figure 10.4 demonstrates generic centralized dynamic navigation. Note that the two-way arrow between the two modules in the mobile unit implies that some implementations may need human assistance or intervention to initialize or reset the vehicle position. Based on the technologies presented in the previous chapters, we know that many variations of this simplified generic architecture are possible. We can add or subtract modules on either the host side or the mobile side to tailor the system in order to meet the user requirements and cost objectives. Because the main emphasis of this book is on the vehicle side of the system, we shall not present a detailed discussion of the centralized approach, but refer interested readers to the references cited in this section for further information.

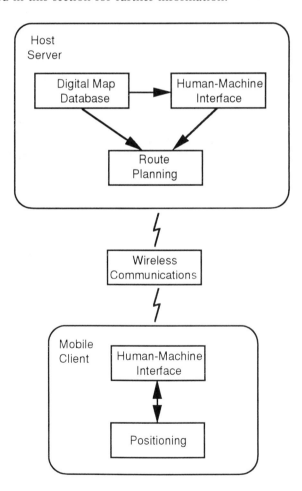

**Figure 10.4** Generic centralized dynamic navigation system.

The portion of the centralized dynamic navigation system involving transmission of navigational information can be implemented using the technologies discussed in Chapter 8. One implementation, EURO-SCOUT, uses a short-range infrared beacon network as the communications medium [7]. All maneuvering instructions and relevant guidance icons displayed on the vehicle are derived using data received from roadside infrared beacons, which are connected via cables to a host in a central facility. If a vehicle equipped with a dead-reckoning positioning module deviates from the planned route, the on-board system switches to autonomous guidance mode, and the system displays a directional arrow pointing toward the destination. Upon encountering the next beacon along the road, the vehicle receives new guidance information and once again returns to its normal centrally assisted guidance mode.

In Japan, the concept of centralized dynamic navigation was first investigated in the 1970s [8,9]. This system was tested in the Comprehensive Automobile Traffic Control System (CACS) project. Unlike recent projects in Japan [10], loop detectors (antenna) at the entrances of intersections were used. A more recent field test, the Universal Traffic Management System (UTMS), utilizes a newly developed short-range beacon network as its vehicle-to-roadside communications medium [11,12].

In the following section, we discuss an approach opposite to the centralized one, that is, an approach in which route guidance intelligence is distributed among all vehicles equipped with navigation units. Unlike the distributed approach, the centralized approach can easily avoid the Braess paradox effect (Section 6.4). The overall efficiency and road utilization should be higher for a road network managed by a centralized host. The total cost of constructing a centralized system with simple in-vehicle units is lower than a distributed system with sophisticated in-vehicle units operating under limited centralized control. On the other hand, the distributed approach does not require a very powerful and robust host computer, reduces the heavy wireless communications load, and allows drivers the freedom to select their own routes if the on-board system provides a choice of criteria for route selection. It also places more of the overall system cost on individual beneficiaries.

## 10.3.2 Distributed Approach

A distributed dynamic navigation system relies on in-vehicle modules for route guidance based on real-time traffic information received from a communications network. In other words, route guidance is solely the responsibility of each individual vehicle.

As in centralized dynamic navigation systems, vehicles equipped with distributed dynamic navigation units receive dynamic information from a host. These systems can also operate as traffic reporting entities (probes) to provide traffic information to the host. Note that other sensors or anecdotal information could be used with or without vehicle probes to provide real-time traffic data. The travel times required to traverse road segments (or links) are recorded by vehicles or other devices. These travel times are then reported to the host via an RF or infrared wireless communications infrastructure (or some other means of communication). The purpose of these reports is to provide real-time traffic information for each link traversed. Given a large enough sample, the current road conditions (travel times) can be statistically determined; these current road conditions can then be broadcast to in-vehicle units so that they can determine proper guidance activities.

One possible method of displaying the road conditions for operators is to use a host-based map display. It can use the traffic congestion information reported by probe vehicles to show traffic conditions in color-coded format. A route in green may indicate free-flowing traffic, whereas a route in yellow might be somewhat congested. Yellow coding could (for example) be used to indicate delay times greater than one standard deviation in the normal travel time. Red could be used to indicate heavily congested, blocked, or travel times two or more standard deviations above the normal travel time. Vehicle travel speeds can also be used for the same purpose. Figure 10.5 represents the recorded speeds for cars traveling along a link with a speed limit of 55 mph (miles per hour) during a 12-hr period of a day. The horizontal axis represents speed, and the vertical axis represents the number of cars at that speed. This relatively small data sample can be used to illustrate this idea (in actual

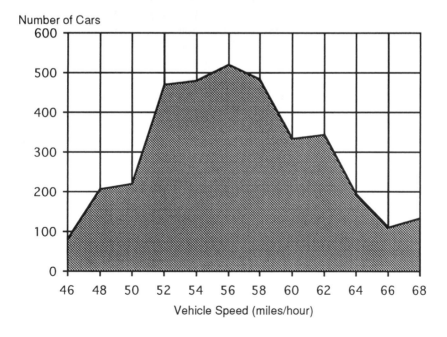

**Figure 10.5** Vehicle speed distribution for a 55-mph speed zone on August 24, 1992.

implementation, a longer period of observations is often required to obtain a sufficiently large set of statistically meaningful data). The mean and standard deviation for these data are 56.56 and 4.38 mph, respectively. Using an approximate value of 5 mph for the standard deviation, this link would be color coded green if the current sample of vehicles was traveling at average speeds greater than 52 mph. A yellow color code would be used if the current sample of vehicles was traveling at average speeds from 47 to 52 mph. Average values below 47 mph would be color coded as red. Similar techniques could also be used in mobile map displays or other centralized location and navigation systems.

The average travel times for vehicles may deviate significantly from the normal travel times for a given time period and road segment. The travel times recorded for the 100 block of one major street for the period Saturday 7:00 A.M.–9:00 A.M. may show no congestion at all. By contrast, the same time period on Monday may show traffic congestion. Given this congestion information, properly equipped vehicles may be diverted to a suitable route around the congestion. Traffic management of this type results in higher utilization of the entire road network while reducing overall network travel times. This helps reduce the need for new and existing roadway construction by smoothing out the peaks during periods of increased traffic congestion.

The wireless communications infrastructure for ordinary vehicles has less stringent throughput requirements. Messages can be queued, and queuing delays on the order of seconds are not as catastrophic as they would be in the case of response times for mission-critical emergency vehicles. This implies a less complex and lower cost message delivery mechanism. On the other hand, a reliable, high-speed message delivery mechanism is a must for any system providing emergency services. This principle applies to all the systems we have discussed.

The basic system architecture for a distributed in-vehicle dynamic navigation system is very similar to that shown in Figure 6.1, plus the two-way communications support and the dynamic route guidance components discussed in Section 6.4. In other words, its architecture is similar to Figure 1.3. The host modules are similar to the diagram shown in Figure 10.4, although the multivehicle route-planning module can be omitted depending on the design. Because in the next chapter we will study a working distributed dynamic navigation system in detail, further discussions of the system organization are left to the following chapter.

Many operational field tests of distributed dynamic navigation are currently under way. The System of Cellular Radio for Traffic Efficiency and Safety (SOCRATES) projects in Europe plan to use the GSM cellular network as a communications medium to test the dynamic navigation concept [13]. The basic one-to-one cellular links are modified to allow the cell to broadcast to as many SOCRATES vehicles as are in the cell at one time. One of the SOCRATES projects, TANGO, uses the Mobitex packet data network (which is equivalent to the RAM network, Section 8.3.5) to communicate between the host and the mobile unit [14,15]. The

Vehicle Information Communications System (VICS) project in Japan uses infrared beacons, microwave beacons, or FM broadcast subcarriers for the communications channels between mobile unit and host [16]; this system has evolved into an operational system (since April 1996), and in-vehicle units are now being sold to motorists. All of these tests have successfully raised public awareness and interest in vehicle navigation for more efficient, safer, and better transportation systems.

Additional applications, such as requests for roadside assistance, accident reporting, and even on-line ordering of tickets for entertainment events as the vehicle approaches the event location, may also be implemented as part of a dynamic navigational system. Many of these applications could be supported using the architectures described in this chapter.

## 10.4 APPLICATIONS: MAYDAY

The automatic or semiautomatic mayday (location-dependent emergency call) system can instantly connect vehicle occupants with a service center for attention or assistance, in addition to automatically reporting the vehicle location. Many people view this as a necessary service. In particular, as noted in Chapter 1, most travelers in the United States rate such a system as their top priority when adding new equipment to their vehicles.

In a related vein, we mentioned E911 at the beginning of this chapter. The advantage of E911 over basic (or wireline) 911 is that the public safety official knows the caller's address and phone number. The U.S. FCC has recently made E911 a mandatory requirement for wireless communications services such as cellular telephone, wideband (broadband) PCS, and geographic area SMR (Chapter 8). This additional requirement and upcoming service is called wireless E911. The FCC requires that within 18 months of the date the rules become effective (October 1, 1996), public safety answering point (PSAP) attendants of wireless communications networks must be able to know a 911 caller's phone number for return calls and the location of the base station or cell site receiving the 911 call, so that calls can be routed to an appropriate PSAP; within 5 years, PSAP attendants must be able to locate the caller within a radius of 125m in 67% of all cases (61 FR 40348-40352). These rules could facilitate the development of many vehicle location and navigation applications that use communications infrastructures similar to those used for wireless E911. As we discuss later, current mayday systems integrate GPS receivers with a communication device to automatic report vehicle location. Once the terrestrial-radio-based positioning technologies improve under the FCC requirement, the GPS receiver would yield to them in many location-related applications. It will certainly have a great impact on future mayday systems.

The services provided by a mayday system could be expanded to include many of the following items:

- Emergency services with location;
- Roadside assistance with location;
- Route assistance or guidance;
- Remote door unlocking;
- Theft detection, notification of theft and stolen-vehicle tracking;
- Airbag deployment notification;
- Travel information (traffic, weather, food, gas, lodging, etc.);
- Hands-free and voice-activated mobile phone or pager.

As with the systems discussed above, mayday systems can take many different forms. For a fleet of service vehicles, a mayday system could be similar to the systems discussed in the last two sections. For a private vehicle, a mayday system is usually simpler in system architecture but more human oriented. In this section, we concentrate on this type of system.

One key feature of the mayday system is its human-centered design combined with a cost-effective location capability. An individual mayday system does not need to communicate with the host on a regular basis (as most AVL systems do), so there is no need for a high communications bandwidth. For increased security and reliability, the system should be activated either by the user or by an emergency event detected by one of the safety sensors. Only at this moment is the communications channel established. It then keeps the user in voice contact with a human operator over the communications channel. The system typically needs to have two-way voice and data communications because the vehicle location information is automatically transmitted to the host once the communications channel is established. As we see in Figure 10.6, the system architecture is much simpler.

Because the mayday system includes a service center operated 24 hours a day and 7 days a week, the host facility may be simplified to a map database with human-machine and wireless communications interfaces. The in-vehicle system may be composed of a microcontroller to control a GPS receiver and a wireless communications device such as a cellular transceiver. As an expandable platform, this system could be augmented to include many other services as described at the beginning of this section. This may require some of the other modules discussed in Part I, as well as safety sensors such as an airbag deployment sensor and a crash sensor.

Figure 10.7 shows the on-board equipment of the mayday system, RESCU (Remote Emergency Satellite Cellular Unit), which is available as an original equipment manufacturer (OEM) product for the 1996 Lincoln Continental. The top box in the figure is a cellular phone transceiver. The bottom box contains a microcontroller, a GPS receiver, a modem, and various control circuits. Model year 1997 Cadillacs have just begun to offer a similar system called OnStar in late 1996.

The RESCU system uses a GPS receiver for positioning and an analog cellular phone (AMPS network, Table 8.3) for communications. It can be activated by either of two buttons in the overhead console. The tow-truck button initiates a roadside

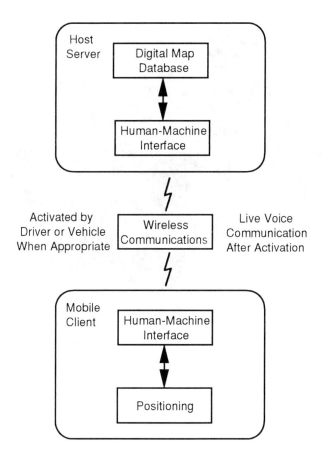

**Figure 10.6** Simplified architecture of a mayday system.

assistance call and the ambulance button initiates an emergency call. Once either button is pressed, a warning light located in the overhead console begins to flash. Status messages are then shown on the multifunction display of the instrument cluster. The cellular phone in the vehicle automatically places a call to the service center. Immediately after the communications channel is established, the system sends the vehicle identification number and location information to the host via the voice channel of the cellular phone using a modem. An operator at the center verifies details of the situation via the cellular phone voice channel and confirms the vehicle location using a map provided by a map database module if possible. If required, the operator may maintain direct voice contact with the driver via a three-way call until help arrives and will notify designated family contacts in case of an emergency. If roadside assistance is requested, the center provides an estimated time of arrival and will call back to confirm that the problem has been resolved.

**Figure 10.7** In-vehicle hardware components of the RESCU mayday system.

Besides the OEM products offered as options on luxury vehicles, other after-market mayday systems either have been available for some time or will be available. The most notable is AutoLink, which also uses cellular for its communications medium and should be on the market very soon. As proposed, in addition to emergency and roadside assistance, one interesting feature of the system is that the vehicle body computer and engine controller are also connected to the host via the communications link. This makes it possible for the host to open door locks and turn off the engine remotely. To request travel information, drivers must use a phone. For instance, when the driver calls for navigation assistance, turn-by-turn navigation instructions are generated by the host and sent to the driver over the communications network. These instructions will be displayed in text form on the screen of the human-machine interface module. Other alternative methods for displaying instructions, such as instructions produced by a control device or voice announcements, are also possible. These proposed features make AutoLink inexpensive, while still providing basic navigational assistance when needed, in addition to the mayday service. The navigation assistance, route guidance, and other travel information features have also been suggested by others to implement over the cellular telephone network [17]. One appealing point of providing these navigation services over these and other types of communications links is that customers pay for the services only when they need them, and there is no need to update individual map databases as required in autonomous navigation systems.

In addition to these commercial products, there are also several ongoing operational field tests of mayday systems. Interested readers can obtain detailed information on these systems from ITS-related conferences or appropriate Internet WWW home pages [18].

Before studying a practical dynamic navigation system in the next chapter, let us think about one question. Suppose that you are responsible for designing a next-generation location or navigation system. Would you prefer placing the intelligence on the mobile unit or on the host? In other words, do you prefer the centralized approach or the distributed approach?

## References

[1] E. N. Skomal, *Automatic Vehicle Locating Systems*, New York: Van Nostrand Reinhold, 1981.

[2] A. Brown and M. A. Sturza, "Vehicle Tracking System Employing Global Positioning System (GPS) Satellites," *U.S. Patent No. 5225842*, July 1993.

[3] A. Brown, J. Siviter, and J. Kiljan, "An Operational Test of a Vehicle Emergency Location Service in Colorado," *Navigation: J. Inst. Navigation*, Vol. 41, No. 4, Winter 1994–1995, pp. 451–462.

[4] C. J. Driscoll, "Finding the Fleet: Vehicle Location Systems and Technologies," *GPS World*, Apr. 1994, pp. 66–70.

[5] Y. Zhao, A. M. Kirson, and L. G. Seymour, "A Configurable Automatic Vehicle Location System," *Proc. First World Congress on Applications of Transport Telematics and Intelligent Vehicle-Highway Systems*, Nov./Dec. 1994, pp. 1569–1576.

[6] M. Sushko, "Vehicle GPS Positioning Over Cellular Networks," *ION Proc. 1994 National Technical Meeting*, Jan. 1994, pp. 261–275.

[7] H. Sodeikat, "Dynamic Route Guidance and Driver Information Services with Infrared Beacon Communication," *Proc. Second World Congress on Intelligent Transport Systems*, Nov. 1995, pp. 622–627.

[8] H. Kawashima, "Overview of Japanese Development and Future Issues," in *Advanced Technology for Road Transport: IVHS and ATT*, I. Catling (Ed.), Norwood, MA: Artech House, 1994, pp. 289–314.

[9] H. Tsumomachi, Y. Miyata, and Y. Kumagai, "RACS and VICS," in *Advanced Technology for Road Transport: IVHS and ATT*, Ian Catling (Ed.), Norwood, MA: Artech House, 1994, pp. 315–330.

[10] N. Yumoto, "Onboard Equipment," in *Advanced Technology for Road Transport: IVHS and ATT*, Ian Catling (Ed.), Norwood, MA: Artech House, 1994, pp. 331–349.

[11] K. Aoyama, "Universal Traffic Management System (UTMS) in Japan," *Proc. IEEE Vehicle Navigation and Information Systems Conference (VNIS '94)*, Aug./Sep. 1994, pp. 619–622.

[12] T. Kitamura, M. Kobayashi, and K. Takeuchi, "The Dynamic Route Guidance Systems of UTMS," *Proc. Second World Congress on Intelligent Transport Systems*, Nov. 1995, pp. 610–615.

[13] I. Catling, "SOCRATES," in *Advanced Technology for Road Transport: IVHS and ATT*, Ian Catling (Ed.), Norwood, MA: Artech House, 1994, pp. 99–118.

[14] P. Geen and T. Biding, "First TANGO in Paris," *Proc. First World Congress on Applications of Transport Telematics and Intelligent Vehicle-Highway Systems*, Nov./Dec. 1994, pp. 2346–2353.

[15] A. Lindkvist, "Evaluation of a Dynamic Route Guidance Field Trial in Sweden," *Proc. Second World Congress on Intelligent Transport Systems*, Nov. 1995, pp. 616–621.

[16] H. Sugimoto, A. Nojima, T. Suzuki, and M. Nakamura, "Development of Toyota In-Vehicle Equipment for The VICS Demonstration Test," *Proc. IEEE Vehicle Navigation and Information Systems Conference (VNIS '94)*, Aug./Sep. 1994, pp. 563–568.

[17]  H. Hakala, K. Vehvilainen, M. Lehto, and S. Turunen, "GSM Based Traveler Information Services Offered with Portable and In-Car Terminals," *Proc. Second World Congress on Intelligent Transport Systems*, Nov. 1995, pp. 1487–1491.

[18]  Intelligent Transportation Society of America, *Access ITS America*, http://www.itsa.org/, 1996.

# CHAPTER 11
▼▼▼

# A CASE STUDY: ADVANCE

## 11.1 INTRODUCTION

ADVANCE is an operational test for the Advanced Driver and Vehicle Advisory Navigation ConcEpt project. It was a test of the dynamic route guidance concept in the northwestern suburbs of Chicago, Illinois. To be more specific, the project used a distributed approach, that is, the intelligence of the dynamic navigation system was placed on the mobile unit. The ADVANCE project is a public and private sector partnership. The main participants in ADVANCE include the Federal Highway Administration (FHWA), the Illinois Department of Transportation (IDOT), Motorola, and the Illinois Universities Transportation Research Consortium (Northwestern University and the University of Illinois at Chicago). In this chapter, we learn how various ideas, hardware, and software (or modules) can be combined to form a successful working system through a case study of this project.

The ADVANCE project was launched in July 1991. A major objective was to determine if motorists supplied with real-time guidance could avoid traffic congestion and improve trip quality. The field test was completed in December 1995 with a targeted deployment (see the next paragraph). The project is expected to be completed by the end of 1996. At that time, all documents concerning this project will become available to the public.

Using a combination of technologies including differential GPS (DGPS), dead-reckoning and map-matching-based positioning, RF wireless communications, dynamic route guidance, CD-ROM map storage, data fusion, and others, the ADVANCE project provided drivers with continuously updated navigation directions and instructions. Although the original project deployment was to have used as many as 5,000 private and test vehicles in a much longer time frame, the recently completed targeted deployment phase had the following features:

- Guidance systems were installed in 80 vehicles.
- Testing was conducted using these existing project test vehicles.
- The "familiar driver" tests using volunteer drivers were conducted for 2 weeks per driver.
- Many aspects of the ADVANCE system were tested in a controlled environment.
- The time frame for testing and evaluation was reduced.

As indicated by the project office, the project gained several benefits under targeted deployment [1]. One of the largest was a financial savings of more than $27 million due to the reduced test scale and time frame. Additionally, the change in emphasis better supported the transformation of the Traffic Information Center into a regional center distributing travel time information for the Gary-Chicago-Milwaukee corridor. Both of these could be accomplished while still testing many of the concepts within the original scope of the study. Another important benefit of the targeted deployment was a quicker turnaround on reports back to the industry, because the overall time frame had been reduced. On the other hand, targeted deployment will have an adverse impact on the statistical significance of some of the data collected and evaluations conducted. We address these issues in the last section of the chapter.

The entire ADVANCE test area covered more than 776.66 $km^2$ (300 square miles), including portions of the City of Chicago and more than 40 communities northwest of Chicago (Figure 11.1). The project included the high-growth areas adjacent to O'Hare International Airport, many offices and retail complexes, road development corridors, and major sports and entertainment complexes. The population in the area is more than 750,000.

The main components of the ADVANCE system were the traffic information center (host) and in-vehicle navigation system (mobile unit) wirelessly connected by a RF communications network. In the following three sections, we discuss each of these components in detail. The chapter concludes with a final section describing the initial evaluation results of the project.

## 11.2 TRAFFIC INFORMATION CENTER

The ADVANCE system consisted of four subsystems: the mobile navigation assistant (MNA), RF communications network (COM), traffic information center (TIC), and

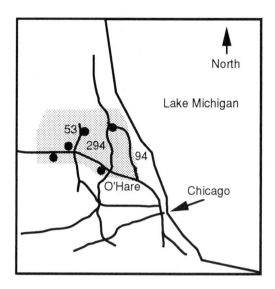

**Figure 11.1** ADVANCE test area.

traffic-related functions (TRF) [2]. These subsystems are shown in Figure 11.2. Because the TRF is closely associated with the TIC, we discuss them together in this section. The MNA and COM are discussed in Sections 11.3 and 11.4, respectively.

The TRF was physically located in the central computer facilities of the TIC and provided transportation-related data and analysis functions using various software processes. A process or task is an abstraction of a running program. In other words, the TRF was a software package that generated both static and dynamic (real-time) traffic information. The static traffic information was developed off-line and was stored in the form of off-line historical profiles. These static profiles were later used by the MNA database compiler to generate route-planning attributes as discussed in Section 6.4. The dynamic traffic information was broadcast to the vehicles on-line in the test area periodically during each predetermined cycle.

To generate the initial static profiles, a large and detailed traffic (equilibrium) flow model was developed [3]. The inputs to this model were the ADVANCE network representation and origin-and-destination (O-D) flow matrices. The static profiles produced by this model were essentially the estimates of initial link travel times, where each link represents a segment with turning delay information.

The ADVANCE network representation was an integrated database for the test area. This database combined elements derived from three sets of data. These data were the NavTech database (Section 2.5.2), the north-northwest network of the Chicago Area Transportation Study (the transportation planning agency for the Chicago area), and traffic signal turning data collected in the field by the staff of

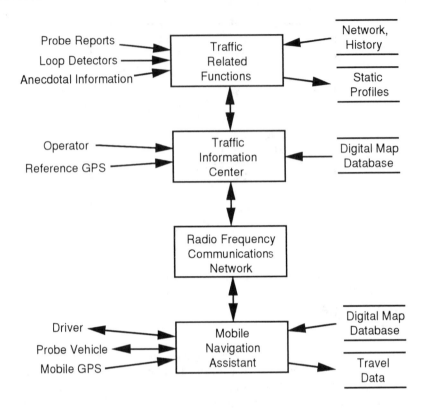

**Figure 11.2** ADVANCE subsystems.

the Urban Transportation Center associated with the University of Illinois at Chicago. Based on these data, all arterial and collector streets were identified, mapped, and coded with respect to street and highway segment attributes. All together, the network representation for the test area consisted of 7,253 directional links. These links intersected in 1,580 intersections. Of these intersections, 699 had signal lights. Note, however, that the map database in the vehicles was significantly larger than the test area. Signals along arterial and collector streets were represented using the link delay functions to be discussed shortly. The test area was defined using 274 boundary links. To represent turning movements, each link included the relevant turning movement. For example, before adding turning movements, one segment was sufficient to represent a road segment approaching a four-way intersection. Once a turning movement is added, it must be divided into a left link, a straight-through link, and a right link. As a result, the total number of links in the network increased by a factor of approximately 2.5.

The O-D flow matrices were tables describing the flow (in units of vehicles per hour) from each possible origin to each possible destination. The origins and destinations were defined in terms of small zones. These zones were about 0.65 to 10.36 km$^2$ (0.25 to 4.0 square miles) in size. All zones were inside the test area. These tables were obtained from the Chicago Area Transportation Study performed in 1990. They were part of the input to the traffic flow (network) model described next.

A traffic flow model was developed in order to obtain initial estimates of link travel times (including turning movements) for various time periods during a 24-hr weekday. The five time periods for a normal weekday are listed in Table 11.1. These estimates (static profiles) were used as initial values for the dynamic route guidance system. The model was a generalized version of the traditional traffic assignment model [4].

Two new features were added to this generalized model. First, a delay function was defined for each turning movement link based on the Australian method [5]. The standard model depends only on the flow along the particular link in question. By contrast, these delay functions determined the appropriate traffic signal control parameters for each intersection in relation to the link flows in the solution. The link delay in this model becomes a function of the flows on conflicting links at the intersection. At intersections without signal lights, the delays are functions of the conflicting flows. In other words, this function incorporates many factors such as number of lanes, left turn pockets, number of phases in the traffic signal, and speed limits. It provides estimates of the average delay per vehicle as a function of the link flow to capacity ratio. Second, the model was solved for the 5 weekday time periods listed in Table 11.1. These five solutions were not correlated. After initial field testing and data collection, these profiles were updated regularly off-line during deployment with actual travel times collected using probe vehicles and other traffic data. Probe vehicles are vehicles that report real-time traffic data for the roads on which they are traveling. The static-profile updates were based on weighted averages of probe data, detector data, and other information discussed later.

We now describe the process used to produce the dynamic traffic information (which was different from the processes described earlier for generating the static profiles) [6]. This process analyzed all incoming data and made incident reports and

Table 11.1
Time Period for Normal Weekdays

| Period | Time |
|---|---|
| Morning peak | 6 A.M. to 9 A.M. |
| Midday | 9 A.M. to 4 P.M. |
| Afternoon peak | 4 P.M. to 6 P.M. |
| Evening | 6 P.M. to 12 A.M. |
| Night | 12 A.M. to 6 A.M. |

predictions about future travel times. The output of these reports was broadcast to vehicles within range of the RF signal of the COM subsystem (module). Vehicles outside the test area approaching or crossing the test area were able to receive dynamic link updates and plan dynamic routes using static information from outside the test area and dynamic information inside the test area. The dynamic traffic data were developed using three tasks of the process: incident detection, data fusion, and travel time prediction (Figure 11.3). These three tasks were executed once per 5-min interval.

The incident detection task provided reports for nonrecurrent congestion. This task consisted of three algorithms. The probe algorithm compared the reported travel time to historical travel time and used the ratio of the two travel times to determine if an incident had occurred. The loop detector algorithm also compared the current traffic volumes and occupancies to historical data to determine if an incident had occurred. (As discussed in Section 8.3.7, loop detectors are magnetic inductive sensors embedded underneath the road pavement to detect the traffic volumes and occupancies.) For detailed discussion of the incident detection based on data provided by loop detectors and probe vehicles, refer to [7]. The raw loop detector data received could be viewed by the operator in a log file window. These data were received from two sources. One was from the Illinois Department of Transportation Traffic Systems Center at 1-min intervals. The other was from the Dundee Road closed-loop signal system at 5-min intervals. The anecdotal information algorithm parsed data received from other facilities. These data included police/fire reports obtained from Northwest Central Dispatch; weather data provided by Surface Systems, Inc.; lane closure data provided by the Illinois State Toll Highway Authority and IDOT; and user reports from *999, which is a cellular-phone-based motorist call-in system. By fusing these data and combining them with the probe and loop detector reports, the fusion task described next decides whether an incident report needs to be issued. Incidents could also be entered manually into the system.

The data fusion task performed three functions in addition to generating incident reports. First, it screened the input data to see if they were consistent and reasonable. For instance, if the probe data were good, they were compared against the loop detector data to see if these loop detector data showed similar results. Second, it converted the loop detector data into travel times by estimating the travel time from the volume and occupancy data. Finally, it estimated the current travel time (fused travel time). If the loop detector was not reliable or malfunctioned, the current travel time was the average of the probe travel times. When both sets of data were available, the following equation was used to estimate the travel time for each road segment:

$$t = \frac{k_p \dfrac{N_p}{\delta_p^2} t_p + k_d \dfrac{w_d}{\delta_p^2} t_d}{k_p \dfrac{N_p}{\delta_p^2} + k_d \dfrac{w_d}{\delta_p^2}}$$

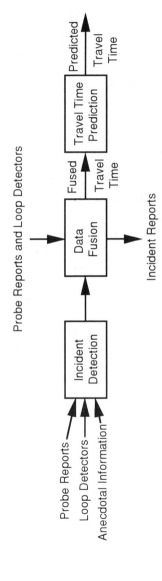

**Figure 11.3**  Real-time data flow in TRF.

where the subscripts $p$ and $d$ indicate the probe and detector data, respectively; $k$ represents the fusion adjustment factor to control the contribution of each piece of data; $\delta$ represents the standard deviation in the corresponding travel time; $N_p$ is the sum of weights for reasonable probe reports; $t_p$ is the mean travel time reported by the probe; $w_d$ represents the weight assigned to the loop-detector travel time; and $t_d$ is the travel time converted from the loop-detector data.

The travel time prediction task used all of the outputs generated using the preceding algorithms to compute travel times for the next four 5-min intervals (these algorithms were executed every 5 min) for 20 min. If an incident occurred, the actual or expected duration of the incident was used as the predicted travel time. Otherwise, the travel times were predicted on the basis of fusion data. Three steps were used. First, the data were aggregated into a single value for the interval. Second, the data were standardized using the mean and standard deviation of the travel time for that link. Finally, the travel times were computed based on the standardized data, link parameters, and other time-related parameters. The final results were in a form ready for conversion to link travel time update messages.

The TIC provided all of the centralized computing resources for ADVANCE. For readers who have not had the opportunity to see a real navigation system and the associated hardware, we will list the sizes of some of the equipment used in this system (Figure 11.4). The hardware components of the TIC were housed in a 178- × 48.4- × 76.2-cm (70- × 19- × 30-in.) standard rack cabinet [6]. The central computer was a SunSPARC 670MP (multiple processors) server with four 70-MHz super-SPARC processors. It had several dual-wide versa module Eurocard (VME) bus slots for expansion. This server had 128 MB of dynamic RAM, 13.6 GB of ROM, a 5.0-GB off-line storage tape drive, and one CD-ROM drive. The server had two 16-port synchronous interface cards that were used for external communications. Additional equipment housed in the cabinet included a modem rack with 16 slots containing 14,400-baud modems. Other stand-alone modems were also used to accommodate modem incompatibilities for leased telephone lines. The main operator consoles were SunSPARC X-Terminals interfaced with the server via 10-BaseT Ethernet. The ac power was supplied through an uninterruptible power system whose internal batteries could provide a 2-hr supply in case of a power outage.

All of the software modules in the TIC interacted with the TIC database and each other using the *interprocess communication* protocol. The interprocess communication protocol is a well-structured non-interrupt-driven communications protocol for software processes [8]. In addition to the TRF processes already discussed, the other software processes or modules supported the user interface and system communications. Like the vehicle databases discussed in previous chapters, the TIC database also contained both static and dynamic data. In addition to the digital map database derived from the NavTech database, it contained static profiles, loop detector reports, probe reports, etc.

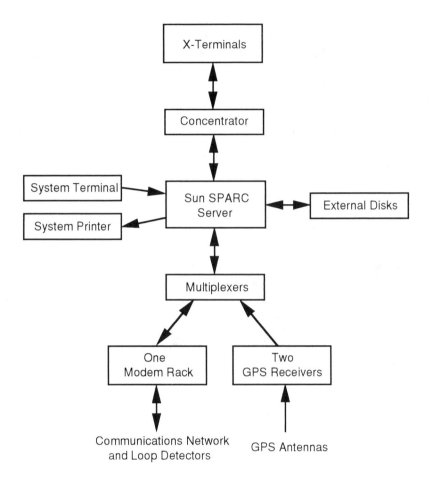

**Figure 11.4** TIC hardware configuration.

The TIC user interface allowed the operator to interact with and control other modules. It could start or stop the entire system, turn various TIC subsystems on or off, monitor the current state of the road network, maintain user account information, and be used to input anecdotal information such as police reports and mobile phone user reports. The display window for monitoring the road network could be used to show the map in different ways via pull-down menus, vertical and horizontal scroll bars, and a mouse. It could be used to select one or more road segments and retrieve information such as street names, city names, zip codes, addresses, travel-time profiles, probe reports, loop detector data, and traffic incidents.

All communications-related tasks were handled by inbound and outbound message processes. The inbound process managed the reception and storage of reports from probe vehicles. Each probe report message was validated on the basis of a unique identifier code. The outbound message transmission via the COM module was handled by the outbound process, a scheduler within the communication module. Each outbound link update message contained a link identifier (ID) and link update information (updated travel time predictions). All the outbound messages were kept in a list in the scheduler which decided when to broadcast these messages. As configured for the targeted deployment, the scheduler allowed for a 4-sec gap between outbound messages to ensure that the COM subsystem did not become overloaded. GPS correction messages [containing DGPS correction data (Section 3.4.2)] were broadcast at 16-sec intervals. Updates of link travel times were transmitted as frequently as the volume of updates permits.

## 11.3  MOBILE NAVIGATION ASSISTANT

The MNA subsystem determined the position of the vehicle using dead-reckoning sensors and DGPS, with assistance from a map-matching algorithm. This module performed route planning based on current traffic information received via the COM subsystem and also provided route guidance instruction to the driver in the forms of both visual display and voice output. Meanwhile, as mentioned, MNA also provided real-time travel time reports for transmission to the TIC. The high-level architecture of this module is identical to that in Figure 1.3. To help visualize the locations of the various system hardware components, Figure 11.5 shows the

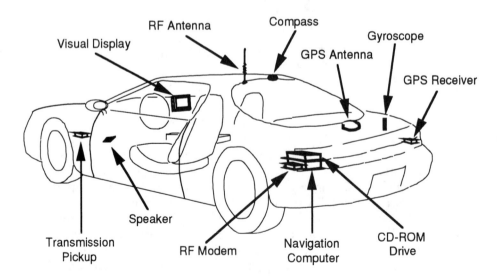

Figure 11.5  In-vehicle equipment for ADVANCE.

in-vehicle equipment used by the MNA. The connecting cables and wires have been omitted for clarity. Figure 11.6 shows the actual in-vehicle MNA display unit.

### 11.3.1 Hardware

The MNA hardware configuration is shown in Figure 11.7. The speaker and display unit with their controller were installed adjacent to the driver as shown in Figure 11.5. A PCMCIA memory card was provided in the display head. The navigation computer, sensor controller, RF modem, GPS receiver, and CD-ROM drive were all mounted in the trunk. The computer booted automatically whenever the ignition was turned on (as one might expect, shut down was performed when the ignition was turned off). A push button under the vehicle dashboard could be used to reboot the computer on demand.

The navigation computer was designed to meet the specifications of the automotive environment, normally −40° to +85°C (−40° to +185°F). In the ADVANCE implementation, the allowable operating temperature range was −30° to +70°C (−22° to +158°F) and allowable storage temperature range was −40° to +85°C (−40° to +185°F). Most parts used for the computer can operate up to +105°C (221°F) to allow temperatures to rise inside the enclosure. The operating humidity limits were 8% to 80% and storage humidity limits were 8% to 95%. As far as vibration was concerned, the computer could withstand a total 10-min duration of sweeping from 15 to 60 Hz in 5 min then back to 15 Hz. As far as shock limits were concerned,

**Figure 11.6** In-vehicle MNA display unit.

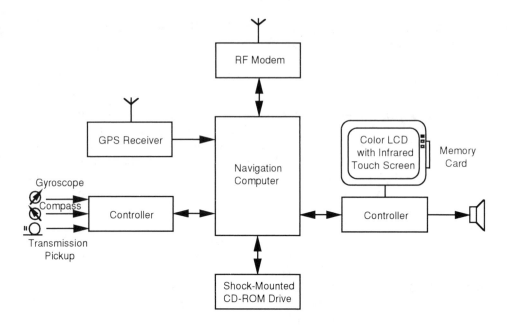

**Figure 11.7** In-vehicle MNA hardware configuration.

the computer could be dropped from a height of 75 cm in a packing carton onto a concrete floor to strike one corner, three edges, or six faces without damage. The operating voltage was from 10.5 to 16V, with a nominal voltage of 14.4V and current of 3A. During startup, it could withstand a voltage of 24V for 2 min.

The touch panel (touch screen) was based on infrared technology. As we saw in Section 7.2.2, this technique has negligible effects on contrast. Evaluations indicated that infrared touch panels had lower glare and better brightness than resistive touch panels in daylight. A backlit color LCD was used as the display unit. The LCD used active-matrix addressing. The size of the screen was 15 cm (5.7 in.) on the diagonal. The size of the display unit was $17.8 \times 12.7 \times 6.4$ cm ($7 \times 5 \times 2.5$ in.). The allowable temperature range was similar to that of the navigation computer. The power supply was 12V at 1A maximum.

The majority of user data entry to the navigation system was performed via the touch panel. Besides the soft (touch) keys on the screen, there were three hard keys on the right side of the monitor that were used for additional controls. The MAIN key was used to display the main screen choices as shown in Figure 11.8. When the car icon on the screen was pressed, the previously stored destination was displayed. If none were stored or the driver wanted to plan a new trip, two choices for method of destination entry would be displayed: by address or by intersection. A keyboard-like touch screen was displayed for entering state names, city names,

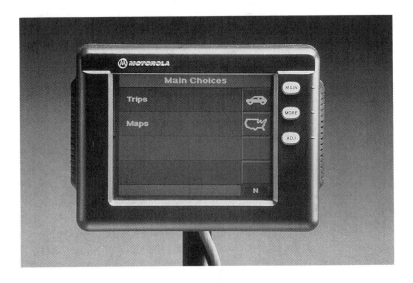

**Figure 11.8** Driver interface.

street names, and addresses. If the map icon was pressed, the road network surrounding the vehicle would be displayed as shown in the route map display later. The "MORE" key was used to display an additional menu, which could be used to enter personal information and show system status information such as the current location. The ADJ key was used to mute or adjust the speaker volume and screen brightness. The current travel direction of the vehicle was displayed using letters (i.e., N, E, NE, NW, etc.) in the lower right corner of each screen. The direction was also indicated by a triangle when the system was in map-display mode.

During the trip, the screen could be in either one of two modes to show the user the current trip status. One display mode used the turn arrow format shown in Figure 11.9. The other mode used the route map format shown in Figure 11.10, which basically consisted of a simplified route map on which the planned route was not highlighted. Recall that we discussed the advantages and disadvantages of these formats in Section 7.2.3. Because most of the touch functions were automatically locked out for safety reasons when the vehicle was in motion, no planning of new routes to unstored destinations could be made until the vehicle was stopped. This user lockout was used to prevent the user from interacting with the interface while driving.

Both display formats showed the name of the road that the vehicle was currently on at the bottom of the screen and the name of the next cross road at the top of the screen. In the turn-arrow format, there was a countdown bar on the left-hand side of the screen with a decreasing mileage number and a shrinking red rectangle. The user could review the summary information for the trip or a list of series of

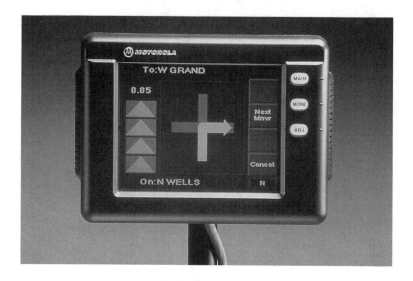

**Figure 11.9**  Turn arrow display.

**Figure 11.10**  Route map display.

upcoming maneuvers at any time by touching the right-hand side of the touch panel. The turn arrow was displayed in yellow and superimposed on the intersection (which was displayed in gray). In the route-map format, a map was displayed to show the area in the immediate vicinity of the vehicle. A red triangle with a white boundary indicated the current location of the vehicle and also the direction of the vehicle as determined by the map-matching-based positioning system. A filled green circle (dot) indicated the location reported by the GPS receiver. If the circle had an orange boundary, this indicated that the DGPS correction was in effect. A yellow dot was used to warn the user that the GPS signal was being received from less than three satellites, with a corresponding reduction in position accuracy. The user could choose also to zoom in or out on the map.

The display monitor could be rotated. Because of the limited viewing angle of the LCD and the need to avoid direct sunlight, this feature was required in order to provide the user with the freedom to rotate the monitor to a comfortable viewing angle.

The MNA used a RAM-based PCMCIA memory card to store traffic data. These data were similar to the data reported back via the COM subsystem. The minor difference was that occasionally a probe report might not be able to reach the TIC. For instance, since U-turns rarely occurred and rarely affected the travel time, this information was not sent to the TIC via the wireless network but kept by the memory card. The information stored in the memory card could be accessed by the TIC later using a card reader to validate and supplement the data received by the TIC via the COM. A 2-MB (formatted to 1.44-MB) static RAM memory card with a memory backup battery was used for this project. It had an operating temperature range of $-30°$ to $+70°C$ ($-22°$ to $+158°F$), the same as the navigation computer.

The CD-ROM drive was shock-mounted on top of the navigation computer as shown in Figure 11.11. The user could change disks by opening the drive cover. The drive had an operating temperature range from $-30°$ to $+60°C$ ($-22°$ to $+140°F$). During the field tests, the upper range became a problem. On some days when the trunk temperature reached $70°C$ ($158°F$), drive performance was erratic since the operating temperature limit had been exceeded. The power supply was 12V. The CD-ROM had a total capacity of 600 MB and its drive had an approximate transfer rate of 150 Kbps. The drive was connected to the navigation computer via a SCSI port.

The positioning sensors were wired to the navigation computer in the trunk. A fluxgate magnetic compass was used to measure the absolute direction of the vehicle. It was either installed in back between the headliner and the outside roof of the vehicle (Figure 11.5) or on the windshield behind the rearview mirror. (Note that the compass should be installed in a location that will minimize abnormal magnetic effects on the measurement.) The power supply voltage range for the compass was 10 to 15V dc, with a typical voltage of 12V. The maximum power supply current was 150 mA. The range of operating temperatures for the compass

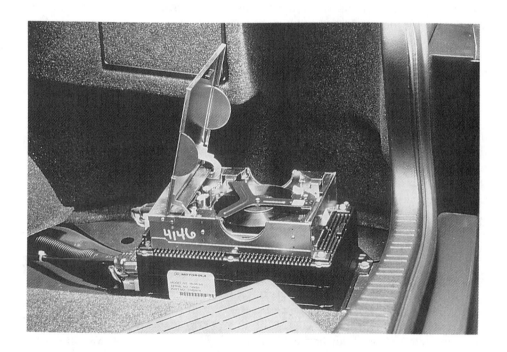

**Figure 11.11** Navigation computer and CD-ROM drive.

was slightly better than that for the navigation computer. A vibration gyroscope with a triangle bar was used to measure the relative direction of the vehicle. It was installed in the side wall of the trunk. The gyroscope must be installed perpendicular to the vehicle's direction of motion with the base glued to a spot with minimum vibration. The gyroscope had a power supply voltage of 8 to 13.5V dc and a maximum power supply current was 15 mA. The gyroscope had an operating temperature range of −20° to +70°C (−4° to +158°F). Distance was measured using transmission pickup sensor. Either a variable-reluctance sensor or a Hall-effect sensor was used, depending on the vehicle model. A variable-reluctance sensor will produce a sine-wave output, whereas a Hall-effect sensor will produce square waves of constant amplitude. A detailed discussion of the operating principles and characteristics of these sensors can be found in Section 3.3.

The GPS receiver and the RF modem were both mounted in the trunk. Their operating temperature ranges were similar to those of the navigation computer and both operated from 9 to 16V dc power supplies. They were both connected to the navigation computer using RS-232 cables. As we saw in Section 3.4.2, the GPS receiver functions as an absolute sensor to report vehicle position, direction, and speed to the navigation system. The RF modem was used to receive the link travel time updates and DGPS correction messages from the TIC and to transmit the current link travel times back to the TIC. This device had a single circuit board which included both a two-way data radio and a modem with a 3W transmitter. To avoid interference, the GPS antenna and the RF antenna were separated by at least 0.914m (3 ft).

### 11.3.2  Software

Because of the real-time and embedded nature of the MNA software, a multitasking real-time operating system was used. To better service the application layer, an operating system services (OSS) layer was constructed on top of the kernel (Figure 11.12). The kernel was the central piece of the operating system that interacted with the computer hardware. The purpose of this OSS layer was to isolate the applications from the host kernel dependencies and provide increased portability to other platforms. The main functions of this layer included the following:

- System initialization and synchronization;
- System shutdown;
- Memory and time management;
- Navigation device driver management;
- Interprocess communication via messaging;
- Mutual exclusion via semaphores;
- Interprocess synchronization via events; and
- Error handling.

Application Layer

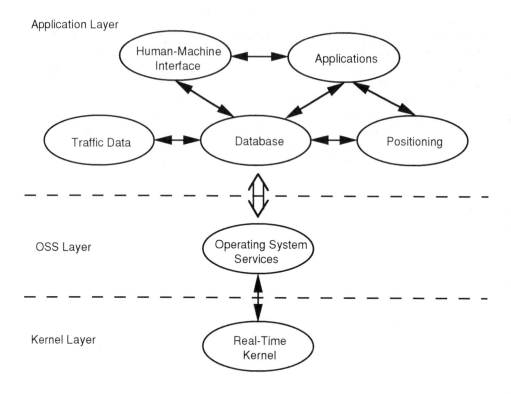

**Figure 11.12**  Main MNA software processes.

The main sequence of execution in this layer was as follows: Create the tasks, set up the memory buffers and disk caches, create the message queues, system semaphores and events, start the tasks, loop and monitor system events, and coordinate system shutdown when required.

The ADVANCE in-vehicle navigation system consisted of all seven basic building modules we learned about in Part I. The software must support the full functionality of these modules. Figure 11.12 shows the main MNA processes. To save space, two-way interprocess communications between processes are indicated using lines with arrows at both ends. A process may consist of more than one task. To avoid cluttering the diagram, a wide double-ended arrow is used to indicate that all the processes in the application layer interact with the layer below, that is, the OSS process. Once the driver has started the vehicle, the navigation computer began initialization and jumped to the execution sequence described in the end of the last paragraph. Once all tasks had started running, the system was ready to accept user input and requests.

Both the structured and object-oriented methodologies were used in the software analysis and design. Because of the variety of tasks involved, this approach provided flexibility to each individual software engineer or team for their own development efforts.

Message passing was the main primitive used for interprocess communication among tasks, with message queues as the communications medium for passing the messages. Two other interprocess communication primitives were also used to support mutual exclusion and synchronization. Mutual exclusion is a way to ensure if one task is using a shared variable or file, the other task will be prevented from doing the same thing [8]. The map data cache is one example of an object that different tasks may want to have read and write access to at the same time. The other two primitives used to support mutual exclusion and synchronization were semaphores and events. Semaphores and events are special variables (or system calls) reserved to report the status of shared variables or files to tasks that need to access them.

We now briefly discuss each process in the application layer. The database process responded to all read and write inquiries from other processes. A hierarchical digital map database was constructed for the MNA. The main reason for using this hierarchical database was to speed up route planning. Because the map data were stored on CD-ROM (which is a device with relatively slow retrieval speed), a cache was implemented in main memory to store frequently accessed data. A database manager was responsible for controlling access to this cache. Different data retrieval inquiries may need to access parcels of data in different layers. Only the route planner was required to have access to parcels in higher layers. All the other modules accessed the lowest layer parcels. As we saw in Part I, this speeds up the time-consuming route-planning process. As defined in Section 2.6.3, a parcel is analogous to a portion of the map area, and is a collection of geographical objects (such as segments) that fall within a rectilinear region defined by minimum and maximum latitudes and longitudes. The dynamic database task received link updates from the TIC and updated the relevant link data stored in main memory (Figure 2.10).

The positioning process included sensor polling, and various integration tasks, as well as a map-matching task. It polled the sensors and fused the resulting data, which were then matched against road segments on the map using a probabilistic method. Once the matching process was complete, the current position and direction of the vehicle were output to the application process and used as inputs for the route-following task and also forwarded to the human-machine process for display on the screen.

The traffic data process was used to log the current link-related travel data. One copy of the log was stored on the memory card. Another copy was sent via RF modem and the COM subsystem to the TIC as a probe-vehicle report. These data (fused with other data) were used to generate predicted travel times for the TIC to broadcast to all test vehicles in the test area. The other data used for fusion here included loop detector data and anecdotal information.

The human-machine interface process was an event-driven process. All other tasks were required to communicate with it via messages. This process first converted the queued message to an event and then processed the event. No blocking was allowed while processing the event. After message conversion, an event queue was used for prioritization. The design and implementation of this process is very important, because it is the one that most affects users' impressions of the system. Figure 11.6, Figures 11.8 through 11.10, and Figure 11.13 show samples of the graphical user interfaces implemented within this process. A detailed discussion of the human-machine interface module can be found in Chapter 7 and the references contained therein.

The application process interacted with other processes and provided many valuable services to the system:

- Planning routes under the shortest time criterion;
- Generating maneuvers for planned routes;
- Supporting maneuver previewing for planned routes;
- Supporting travel distance display;
- Guiding the vehicle along a planned route;
- Supporting on-demand off-route recovery;
- Using dynamic information received from the TIC to replan routes;
- Forwarding current vehicle location and motion status to the human-machine interface.

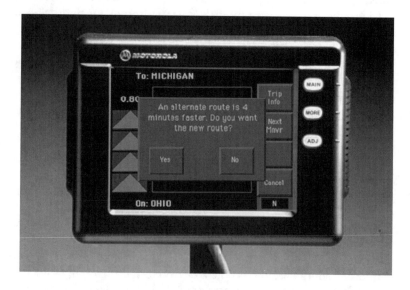

**Figure 11.13** Typical display during dynamic navigation.

All of these services were provided by different tasks. The two most important tasks were route planning and guidance. The planning task was based on a bidirectional hierarchical heuristic search algorithm. The evaluation (cost) function of the algorithm was based on the travel time. During route planning, an update message was displayed on the screen to indicate what percentage of the planning had been completed. Once the route was planned, the route segments were then passed to the route guidance module. First, each higher layer segment was broken into segments at the lowest level, and maneuvering information such as turn types and maneuver types was added (Figure 6.2). The result was a maneuver list that the route-following task could follow using the position information provided by the positioning process. This module also periodically sent messages to the human-machine interface module to identify the next set of instructions to be fetched from CD-ROM to main memory. This allowed the human-machine interface to provide turn-by-turn prerecorded human voice instructions to guide the driver. One important feature of this process was that a better route could be planned whenever the route-planning module discovered that a link cost on the planned route had been updated. The route-planning module replanned the route based on the newly updated link travel time. After completion, it would compare the new route against the original route to see if the new route was better. If the new route had at least a 2-min time savings compared to the original one, it would pop up a window on the display unit and unlock the screen to ask if the user wanted to take this new route. Both the visual display and voice were activated to remind the user. Figure 11.13 shows a typical dialog box. If the user accepted the new route, the original route was deleted and all route guidance activities were transferred to the new route.

One interesting challenge that the project faced was how to make the road segments used by the MNA compatible with the links used for travel time updates by the TIC. The MNA static database represented the road network using nodes and segments (recall that these terms can be found in Chapter 2). In contrast, the TIC used links to represent the dynamic travel time information associated with the corresponding segment (recall from Chapter 6 that a link is a segment plus a turning movement). The link cost is the travel cost of the corresponding segment plus a turning cost. If we used geographical coordinates (latitude and longitude) to match links with segments, the message would be much longer, and RF capacity would be wasted. A simple way to solve this problem is to assign a unique identifier (ID) to each link and associated segment. One naming algorithm first assigned sequential IDs to all nodes in the test area following a predetermined assignment pattern [9] (recall that a segment is a piece of roadway between two nodes). After the node assignment, the segment could be represented by a unique pair of IDs identifying the end nodes of the segment. The algorithm then assigned unique IDs to each link to match pairs of segment IDs. These unique IDs were used to identify links in communications messages. After these messages were received, the MNA then used these link IDs to locate the relevant segments for updating the dynamic database.

Meanwhile, the MNA transmitted probe report messages to the TIC. These probe messages included link descriptions and link costs converted from the corresponding segment IDs and travel times.

The reader may have already noticed that the combinational effects of the processes discussed form a well-structured dynamic navigation system functionally equivalent to that shown in Figure 1.3 of the introductory chapter. As mentioned in Section 10.3.2, the various functions implemented in the application layer are a vivid result of integrating the basic modules (building blocks) we studied in Part I. Once the individual location and navigation modules have been completed, it will be much easier to use these modules or variations on these modules to implement a completely new, stand-alone system.

## 11.4 COMMUNICATIONS NETWORK

Since the communications network (COM subsystem) was a significant component of the ADVANCE project, design requirements were established early in the development that:

- Restricted communications development to a minimum;
- Covered the defined test area;
- Provided communications for a fleet of 5,000 vehicles (later reduced to 80 vehicles for targeted deployment);
- Provided two-way communications;
- Provided for staged growth;
- Provided messaging that was not critical to public safety;
- Provided real-time communications;
- Supported point-to-point inbound and broadcast outbound communications.

Based on these requirements and the available technologies, the full system was constructed using a phased development methodology.

The Motorola Data Communication (MDC 4800) system was selected for the COM subsystem used during the targeted deployment phase of the project. This is a private land mobile radio system using a central dispatch approach. It consisted of an infrastructure at a central location that provided two-way communications for participating vehicles. Each of the probe vehicles could initiate or respond to the central station. The system configuration is shown in Figure 11.14, where P-A stands for preamplifier and M stands for modem. The COM subsystem and TIC could be physically located at the same site or at two separate sites, as was the case for the ADVANCE system.

The COM site was housed in a 59.44m (195 ft) high leased facility in the south central portion of the test area. This site was chosen because RF coverage

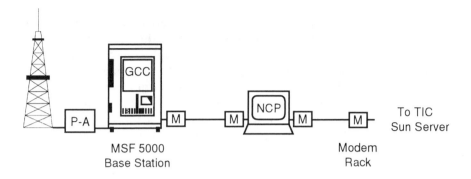

**Figure 11.14** RF COM infrastructure.

prediction studies indicated it provided the best coverage to the ADVANCE test area. The site consisted of an MSF 5000 800-MHz base station, a General Communications Controller (GCC), and a preamplifier transmitting on 853.0625 MHz into 12-dB gain for outbound messages and receiving on 808.0625 MHz transmitted at 3W into 3-dB gain for inbound messages. The GCC is a processor that encodes and decodes various communications protocols. The base station and GCC were housed in a 122- × 55.88- × 58.42-cm (48- × 22- × 23-in.) cabinet and the preamplifier was housed in a 45.72- × 7.62- × 20.32-cm (18- × 3- × 8-in.) cabinet. The base station can support three channels. For this field deployment, only one channel was used for both transmitting and receiving. It was connected (via the preamplifier) to the antennas with 56.4m (185-ft) runs of 4.13-cm (1-5/8-in.) diameter Heliax. The antennas (DB810K-XT) were omnidirectional and had 10-dB gain. Through modem connections, data could be transmitted over a leased telephone line to the NCP (network control processor) at the TIC. The NCP can route messages for up to 16 base stations. Therefore, the NCP could support a maximum of 48 channels. Further evaluation concluded that the test area had good data communications coverage [10]. The actual coverage was 99% (which was higher than the 90% design requirement).

For efficient utilization of resources, the one-on-one transactional messaging was eliminated for the outbound channels, while a one-to-many broadcast approach was used by the base station. By broadcasting the outbound messages, message acknowledgments and individual addressing were eliminated from the communications protocol. Identical messages were broadcast on the outbound channels, while individual inbound messages were transmitted on the inbound channel from each probe vehicle to the base station. A one-to-one transactional messaging protocol was still used for these transmissions. Probe vehicles were divided into groups locked onto a given pair of inbound and outbound channels. No roaming was allowed among inbound channels, in order to facilitate staged growth of the system. All messages were encoded in binary form, with a maximum of 240 bytes for each outbound message and a maximum of 50 bytes for each inbound message. This shorter, 50-byte message length increased the possibility for each probe message to reach the COM subsystem. Two inbound-message throttling methods were designed in order to prevent runaway lengths and to ensure that communications could be maintained under severe overload conditions. To further limit the communications traffic, travel time information for residential streets was not sent over the air. The maximum number of retransmissions for inbound messages to the TIC was limited to four. Because the base station had a power of 100W and mobile transceiver power had a power of 3W, every received signal was amplified 10 dB before being processed in the base station.

The COM subsystem was connected to a modem in the TIC server modem rack over a leased telephone line via an NCP as shown in Figure 11.14. The leased line had a transmission rate of 9600 bps. As we mentioned earlier, the COM subsystem handled both the inbound and outbound data flow. The travel time predictions and other pertinent data generated by the TRF were sent to the probe vehicles via the COM subsystem. In return, the COM subsystem received current travel times from each of the probe vehicles. The implementation of the communications interface protocol conceptually followed the ISO OSI Reference Model shown in Figure 8.10.

With the broadcast architecture, binary message encoding, inbound message throttling, and reduced number of link-update messages, 20 MDC 4800 RF channels would be required for a fleet of 5,000 vehicles. This was not feasible under the architecture discussed previously. An alternative 19.2-Kbps protocol (RD-LAP; see Table 8.6) was investigated to replace the MDC-4800 protocol; this would have required no changes in the TIC and MNA. This alternative would have reduced the number of RF channels from 20 to 3. Because there were no RD-LAP modems available to accommodate the ADVANCE project schedule, other technologies were evaluated such as ARDIS (Section 8.3.5) and broadcast subcarriers (Section 8.3.6). Eventually, a COM design that could support up to 5,000 vehicles in the area (for later phases in the project) was completed. The new system was based on the iDEN 64-Kbps communication system (Section 8.3.4). The most important feature of the iDEN option compared to the RD-LAP was the higher data rate, so that only two RF channels were required for the 5,000 probe vehicles. Due to the targeted deployment, this version of the COM was never deployed into the field, although operational test results indicated that this system worked extremely well.

A great deal of useful experience was obtained during implementation of the COM subsystem. For instance, most commercially available one-way or two-way RF communication systems are optimized for a high number of individual transactions. By contrast, in the COM architecture, outbound messages were broadcast continuously whereas inbound transactions occurred only occasionally. This unique architecture required modifications of available off-the-shelf equipment. Furthermore, commercial equipment is often designed for certain radio bands. Unfortunately, these bands are saturated in most metropolitan areas. Therefore, careful evaluation must be made before selecting a suitable RF transmission product.

ADVANCE was successful in achieving the major field-test objectives established in the early 1990s. During the course of project development, many challenges were encountered [11]. However, many useful lessons were learned. Some late requirements added to support operational field testing significantly affected system design, and resulted in the system discussed in this chapter. One example was the decision to record detailed traffic data for postevaluation. The earlier design had to be changed or modified along the way in order to incorporate new hardware and software for this additional feature. Despite incremental changes for tailoring the system to satisfy many different expectations proposed by all interested parties, the major goals were achieved, and deployment and evaluation provided many valuable results.

## 11.5 INITIAL EVALUATION RESULTS

As of this writing, final evaluations of the ADVANCE project are still under way. We first discuss some of the initial findings and then direct readers to references for the upcoming final results at the end of this section. Before turning our attentions to these findings, we first consider the major evaluation objectives of this project, which will help explain the origin of these and other findings. The evaluation included the innovative features of the system, in particular the concept of vehicles as traffic probes. It emphasized the following:

- Evaluation of TIC for meeting the present and future needs of the partners involved in the project;
- Assessment of TRF and probe vehicles versus loop detectors for collecting real-time traffic data;
- Determination of whether in-vehicle route guidance can be beneficial in avoiding urban congestion for familiar drivers;
- Examination of the effectiveness of dynamic route guidance in reducing travel times;
- Summarizing of the achievements and lessons learned;
- Review of institutional arrangements for future efforts.

Because the targeted deployment involved fewer probe vehicles than originally planned, the field tests were conducted in a more controlled environment. This enabled the participating parties to evaluate well-defined characteristics of the system more accurately. On the other hand, this approach might also hinder the ability to test the robustness of the system to handle uncontrolled random and emergency situations as well as much larger fleet dynamics. In addition, the data collected by the probe vehicles were not sufficient to model the traffic flow over the entire road network in the test area.

Before drawing any conclusions from the initial findings discussed here, we would like to remind the reader that all of these results were based on one particular system tested in one particular test area under certain restrictions. For instance, the system did not provide any traffic information for local streets. The road network in the test area is generally a regular rectilinear arterial grid, which makes it difficult to provide far more efficient alternative routes. Test drivers were familiar with the road network and voluntarily participated in the evaluation. During the evaluation, it was discovered that the system was effective in reporting traffic emergencies and directing drivers to alternative roads. This was particularly helpful for highway travelers. Incremental improvements were made in the dynamic database, network model, and historical profiles (travel time files) during the deployment period.

The study found that reasonably accurate travel time predictions for the upcoming 15 min on a link can be provided by at least three probe vehicles that have just traversed that link over the preceding 15-min period. Because the loop detectors were placed to operate a "closed-loop" signal system, information from these detectors was of marginal value in traffic congestion measurement. Probe reports provided more accurate estimates of travel time. These insights were based on more than 50,000 probe reports, of which 99.4% were reliable.

Nearly 100 origin-and-destination runs were conducted to compare the performance of static and dynamic navigation. Test vehicles were divided into two groups traveling on the same route: One without real-time traffic data and the other with real-time traffic data. Only occasionally were large time savings found. On the other hand, the number of test runs available on any particular route was generally too small to obtain statistical verification of significance.

The evaluation team found no strong evidence from the available test runs that real-time traffic data could make a significant impact in reduction of users' travel time and improvement of performance on arterial roads. The reason may be that familiar drivers are often able to avoid traffic congestion or trouble spots by driving on alternative roads, frequently on residential streets, or by relying on radio traffic reports. In contrast, community standards may be infringed by excessive routing to residential streets; this was reflected in the ADVANCE route planner. Another possible reason is that traffic flows on arterials are mainly determined by turn signal timing. The latency of real-time traffic data arriving at the vehicle is not so large as to affect dynamic route guidance. The turning movement information included for guidance did not seem to lead to any noticeable decrease in travel time or have any significant impact on the system as a whole. There was no significant difference in time savings from incorporating the delays for left turns, straight-through turns, and right turns. The system was effective at saving the traveler's time on highways, perhaps because there are only a few entry and exit ramps on these long roads. Similarly, the unimpeded nature of highway traffic allowed the collected data to represent the highway network more accurately than the network of arterial roads.

The ADVANCE operational field test focused on drivers who were familiar with the road network and communities in the test area. The initial findings from the survey responses to the dynamic route guidance systems led the evaluation team to conclude that they had different needs and desires than unfamiliar drivers, such as visitors, who may be more concerned with obtaining accurate directions than saving travel time [12]. Familiar drivers expect a more intelligent navigation computer that can incorporate their knowledge or learn from their experience, rather than simply being told where to go by a predetermined algorithm. This finding was obtained from a small fleet driven by familiar drivers with very little real-time traffic data available to support the dynamic route guidance, because the entire test field had at most 80 equipped vehicles in addition to the loop detectors and anecdotal information. The 2-week test period was relatively short. All ADVANCE test drivers were volunteer members of households [13]. Most households had incomes of more than $50,000 per year and had two or more motor vehicles. The mean age of the drivers was 44 years, and 62% of them were males. This sample is biased compared to a random sample of the population.

A focus group study of 32 ADVANCE test drivers found that drivers were more interested in dynamically updated traffic information than a route planner that uses static information. Database errors and the fixed criteria for route selection and planning meant that the quality of planned routes was not sufficiently high. Participants wanted more control over the planning process and the flexibility to choose planning criteria. Some preferred to be presented with more than one route. Participants suggested that a route planner probably should learn from the daily routes taken by the driver. Real-time traffic data could be used to evaluate these routes so that faster alternatives could be generated if necessary. Participants believed that drivers would be willing to pay for future real-time route guidance systems and services. Some were willing to pay from $750 to $1,000 for an in-vehicle unit in addition to from $2 to $30 per month for "single-area" coverage of real-time traffic data.

One hundred and ten drivers returned the follow-up survey after the field test. In the survey, they rated the importance and performance of 13 system features. Destination accuracy, route quality, system reliability, route-planning performance, and tracking accuracy received the highest scores for importance of features (Figure 11.15). Of these, route quality and route-planning performance were rated the features with lowest performance. Responses to the willingness-to-pay-for-future-route-guidance-systems question were in the range of $500 to $2,000, with mean price of $980. All these findings suggest that a more intelligent route planner may be required in order for in-vehicle guidance systems to attract familiar drivers to the market. It has been speculated that U.S. car rental companies may be the largest buyers of static navigation systems at this point because most rental car drivers are

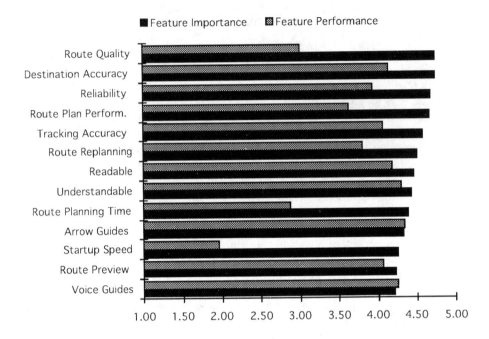

**Figure 11.15** Familiar drivers' ratings of ADVANCE guidance features (*Source:* ADVANCE project.)

unfamiliar with the area in which they are driving and their priority is to follow well-presented routes to reach their destinations.

Questions remain with respect to the cost of the two-way communications equipment and airtime. This question must be answered before massive deployment of this type of system can occur. Periodically reporting traffic data back to the TIC takes significant airtime. If the vehicle owner were responsible for the entire payment, few would be willing to purchase this communications equipment.

The limited evaluation results seem to suggest that dynamic route guidance will enjoy only mixed success without instant real-time traffic data updates. It is very likely that a successful system will have to be able to detect and respond to large and unexpected delays on the network and provide more intelligent assistance to drivers. Because small savings in travel time may not be enough to convince familiar drivers to follow the guidance suggestions and to purchase the unit, more studies and research must be done to ensure that such systems can meet the expectations of the general public, especially familiar drivers.

Interested readers can obtain more information about this project from the ADVANCE Internet WWW page [14], which points to all available technical documents, working papers, and reports. A series of white papers should be available there by the time this book is published so that readers can study final evaluation results.

Let us consider one question to conclude this chapter. The ADVANCE system was designed in the early 1990s. With the location and navigation modules you learned and the new technologies available right now, could you design a dynamic navigation system that achieves the same functionalities as ADVANCE, but with much more efficient performance and cost-effective components? More simply, can you improve any one of the subsystems or components described in this chapter?

## References

[1] J. F. Ligas and S. Bowcott, "ADVANCE: Initial Deployment," *Proc. 1995 Annual Meeting of ITS America*, Mar. 1995, pp. 291–296.

[2] D. E. Boyce, A. M. Kirson, and J. L. Schofer, "ADVANCE—The Illinois Dynamic Navigation and Route Guidance Demonstration Program," in *Advanced Technology for Road Transport: IVHS and ATT*, Ian Catling (Ed.), Norwood, MA: Artech House, 1994, pp. 247–270.

[3] D. E. Boyce, A. Tarko, S. Berka, and Y. Zhang, "Estimation of Link Travel Times with a Large-Scale Network Flow Model for a Dynamic Route Guidance System," *Proc. IVHS America 1994 Annual Meeting*, Apr. 1994, pp. 37–43.

[4] A. Nagurney, *Network Economics—A Variational Inequality Approach*, Boston: Kluwer Academic, 1993.

[5] R. Akcelik, "The Highway Capacity Manual Formula for Signalized Intersections," *ITE J.*, Vol. 58, No. 3, 1988, pp. 23–27.

[6] J. Dillenburg, C. Lain, P. C. Nelson, and D. Rorem, "Design of the ADVANCE Traffic Information Center," *Proc. 1995 Annual Meeting of ITS America*, Mar. 1995, pp. 321–327.

[7] V. Sethi, N. Bhandari, F. S. Koppelman, and J. L. Schofer, "Arterial Incident Detection Using Fixed Detector and Probe Vehicle Data," *Transportation Res. C*, Vol. 3, No. 3, Apr. 1995, pp. 99–112.

[8] A. S. Tanenbaum, *Modern Operating Systems*, Englewood Cliffs, NJ: Prentice Hall, 1992.

[9] P. C. Nelson, P. Petrov, and P. Pollock, "Assigning Segment and Link Identifiers for ADVANCE," *Proc. Intelligent Vehicles Symposium*, July 1993, pp. 370–372.

[10] R. Makaras, *RF Data Communication Coverage Test Report—4.8 Kbps MDC 4800 Protocol*, Motorola, July 1993.

[11] J. F. Ligas and S. Bowcott, "Implementation of ADVANCE," *Proc. 1996 Annual Meeting of ITS America*, Apr. 1996, pp. 87–92.

[12] V. W. Inman, R. N. Fleischman, R. R. Sanchez, C. L. Porter, L. A. Thelen, and G. Golembiewski, *TravTek Evaluation: Rental and Local User Study*, McLean, VA: U.S. Dept. of Transportation, Federal Highway Administration, Report No. FHWA-RD-96-028, Mar. 1996.

[13] J. L. Schofer, F. S. Koppelman, and W. A. Charlton, "Familiar Driver Response to In-Vehicle Route Guidance Systems," *Abs. Third World Congress on Intelligent Transport Systems,* Paper #1112 (full paper in CD-ROM), Oct. 1996, p. 82.

[14] Argonne National Laboratory, *Advanced Driver and Vehicle Advisory Navigation Concept*, http://beijing.dis.anl.gov/ADVANCE/, 1996.

# CHAPTER 12
▼▼▼

# CONCLUSIONS

As we have seen, development of a location and navigation system is not a simple task. However, with the broad background and technologies studied so far, we should be able to make this task much easier. In other words, we need to properly identify system requirements, determine the system architecture, identify the functions to be supported, specify the corresponding modules, and select suitable hardware components and software tools to design and implement these modules creatively based on the principles learned in this book. Furthermore, we need to test and integrate these modules into a flexible, expandable, and compatible system. We have learned that a variety of modules can be implemented (Part I) and different systems can be constructed on the basis of these modules (Part II). In particular, studying Part II has provided us a better appreciation of the intricacies of systems engineering. As we mentioned at the beginning of this book and will discuss in a later section here, without proper system-level design, systems may have deficiencies that can be difficult to correct even if the modules are individually functional. We conclude this book with a discussion of lessons learned from past experience and possible future directions for vehicle location and navigation systems.

## 12.1 PAST LESSONS

The discussion in this section focuses more on engineering-related than on other issues such as funding, marketing, policy, legal, and private-public partnerships, which have also arisen in the past. Interested readers can find discussions of those types of issues in the proceedings of conferences organized by European Road Transport Telematics Implementation Coordination Organization (ERTICO), Intelligent Transportation Society (ITS) of America, and VERTIS (Vehicle, Road and Traffic Intelligence Society, Japan).

Good market forecasts are required for development of any location and navigation product. Different markets may have different cultural backgrounds, environmental issues, and different types of driving (rental, commuting, shopping, etc.). Compared with the Japanese in-vehicle navigation market, the European and U.S. markets are still in the early stages of development. As stated in Chapter 1, according to Japan Electronic Industries Association, sales of vehicle navigation systems to consumers in Japan reached 530,000 units in 1995, up 73.2% from 1994. More than 30 companies are competing in a market that has seen prices of some units drop from U.S. $6000 eight years ago to around $1900 today. In Japan, addresses are based on land sectors or on the order of construction, rather than location on a street. Only major streets are named, and most small streets have no names. Therefore, map drawing and reading are essential skills for survival in Japan. Directions are often given using major landmarks on maps rather than street names, traffic lights, and stop signs. With this cultural background, it is not difficult to understand why the map-display-based navigation system has been very well embraced on the Japanese market. The method of navigation in Japan is to superimpose a route on the map and guide the driver to an area close to the destination. On the other hand, in the United States and in European countries, every street has a name and every building or house has an address, with the special case that many major European cities have quite a few traffic circles. The navigation usually relies on turn-by-turn instructions to guide the driver to a specific address. Furthermore, in the United States, people are more concerned about their personal safety and the safety of their vehicles and less worried about finding a location (because of their addressing system is relatively easy). These cultural and environmental differences imply quite different attitudes toward navigation products.

In the United States or Europe, a vehicle navigation system often provides turn-by-turn guidance. It is important for such a system to have a large main memory, a fast database access storage device, and a powerful CPU. Even with a fast, large memory and a high-performance CPU, better data structures and algorithms are still required to increase the speed of the system. On the other hand, powerful graphics processing is essential for the rapid map-scrolling and scale-change features found in Japanese navigation systems. Although the distinction may not be very clear, map-display-oriented navigation systems often need more graphics processing power while turn-by-turn-oriented navigation systems require rapid route calculation and real-time guidance response. Without any question, a product would be successful on the U.S. and European markets if it could be developed to maintain a low selling price and comfortable profit while providing both sets of features. However, this is easier said than done. Products should probably first target a specific market segment.

A stable and accurate map database is required to develop a quality navigation system. This database can help improve system accuracy and reduce subsequent modification of the navigation software. Many features of the display and traveler information (human-machine interface), positioning, map-matching, route-planning, and route guidance modules rely on database attributes, including road type, maneuver type, turn type, street speed, street name, etc. Any database changes could result in a significant reengineering effort to rebuild the navigation system.

Project managers or lead engineers must be willing to take a concrete step-by-step approach starting from the basics, to have good interpersonal skills and a broad background, to learn new technologies, and to have faith and stick with the project in order for it to be successful. Any work on a new subsystem, prototype, or product should start with the basics. In other words, it is better to make it as simple as possible but make sure it works correctly and is expandable from the very beginning. Any new features or enhancements can then be easily added to a working system. As discussed in the beginning of Part II, a good system architecture is essential to providing a stable basis for any expandable and flexible system. For a complex product that involves both hardware and software, there must be an experienced systems engineer (architect) or a team of systems engineers to coordinate design and development. Otherwise, costly testing and integration as well as temporary patches could mean more trouble later while trying to push a product out the door for a delivery deadline. However, even a good system architecture is normally only suitable for a limited range of system sizes. A completely new kind of problem should be expected every time the system is enlarged by a factor of two. Software engineers should be involved from the beginning of the design phase to avoid the necessity of tailoring code to already finished hardware. It is always worthwhile to spend more time on the system specifications. Finally, the managers and engineers involved in any location and navigation project must keep faith in the project, which is equally important to learning new technologies. A firm belief in the project, persistent effort, and full support from upper-level management, especially during difficult times, are critical for any success.

A thorough understanding of software project management and the development life cycle is particularly important because there are so many critical programs embedded in the location and navigation system. Software project management is also an engineering discipline in its own right, with predictable patterns. Many experts believe that a software organization or development team should use the capability maturity model (CMM) developed by the Software Engineering Institute (SEI) to measure the maturity of the organization [1]. SEI is a federally funded research and development center that is affiliated with Carnegie Mellon University in the United States. The higher the SEI maturity level a software organization has, the better a software product it can produce. Even using the CMM, many of the system-level issues involved in a complicated embedded multitasking system require experienced and dedicated engineers.

Configuration management should be adopted at the very beginning of the project. Configuration management is a process to maintain the integrity of products as these products evolve through the development and production cycles [2]. It identifies, controls, and provides the status of defined hardware, software, and firmware end items. Software configuration management is especially useful for mission-critical software development projects involving teams of developers. Through documenting and tracking software during its life cycle, software configuration management ensures that all changes are recorded and that the current state of the software is known and is reproducible [3]. In addition, all project-related documents should be placed under configuration management. After all, this is a vital component of product quality management and a key to maintaining competitiveness in today's global marketplace.

When selecting software methodologies, both managers and engineers must understand the pros and cons of the various methodologies available. In particular, correct assessment of available methodologies is very important for any real-time embedded application. Structured design and object-oriented design are the two most popular methodologies. Both of them can be used in embedded systems. However, when a particular language is employed for implementation, we must know the consequence of applying it on a dedicated hardware target. This is especially important for an organization where most of the programmers are familiar with a structured language and have just begun the transition to an object-oriented language. For instance, C is a general-purpose programming language that is relatively compact and efficient. On the other hand, C++ is a "better" C, which supports better data abstraction and object-oriented programming. However, C++ is a far more complex language than C, run-time supports are larger, and run-time costs are often hidden. C++ compiler diagnostics may be difficult to understand and run-time bugs may be subtle. Improper implementations of virtual functions in C++ may increase object size and slow down function calls. Constructors and destructors in C++ (analogous to malloc and free in C) may add overhead to object creation and destruction, and make interrupts and exception handling more complicated. Classes in C++ may have hidden fields that make them difficult to map into existing data structures. Temporary objects in C++ may take unexpected time and space during execution. There could be memory leak problems if no special attention is paid to the memory usage of each programmer or team. Given experienced and properly trained engineers, all these problems may be avoidable.

There is also a wide variety of emerging technologies. Recent references are available from which a reader can learn these technologies and proven theories and learn how to more efficiently implement real-time embedded software. Rate-monotonic scheduling theory is a useful tool that allows system developers to meet the requirements of real-time systems by managing system concurrence and timing constraints at the tasking and message-passing level [4]. It ensures that all tasks meet their deadlines as long as the system utilization of all tasks lies below a certain bound and appropriate scheduling algorithms are used. System designers can use this theory to predict whether task deadlines will be met, long before the expensive implementation phase of a project begins.

Specification and Description Language (SDL) has been standardized by the International Telecommunication Union (ITU, formerly CCITT) for specifying the behavior of real-time communications systems [5]. The behavior descriptions are based on communicating extended state machines that are represented by processes. An extended state machine consists of a number of states and a number of transitions connecting the states. One of the states is designated the initial state. Communication is represented by signals and can take place between the processes themselves or between the processes and the environment containing the system model. Aspects of communication between processes are closely related to the description of system structure. Probably less known in the ITS field, SDL is widely used in the telecommunications field. SDL is not just for describing telecommunications services. It is a general-purpose description language for communications systems that have complex event-driven, real-time, and interactive applications involving many concurrent activities. Recall that a communications module is a vital component to improve performance and increased functionality of vehicle location and navigation systems as discussed in Chapter 8. SDL could be a very useful development tool for large and complex ITS communications.

Well-structured object-oriented software architectures have numerous patterns. These patterns describe simple and elegant solutions to specific problems in software design. Design patterns capture reusable designs that have developed and evolved over time. Twenty-three very useful design patterns summarized from decades of object-oriented software experience can be found in [6]. For everyday programming, sixty-one short "heuristics" (rules) are presented in [7]. They act as guidelines that can easily be memorized and used in object-oriented design. As mentioned, the object-oriented methodology can be used for embedded real-time system development, but care must be taken when using the C++ programming language because many programmers are migrating from the more familiar C language. Embedded software is tightly connected to its physical environment through various human interventions, sensors, and actuators. It must respond to many events within specified time limits. Interested readers will find additional information on object-oriented methodology and real-time systems in [8], which discusses a variety of detailed design issues and implementation aspects specific to embedded real-time systems when using the C++ programming language.

## 12.2 FUTURE DIRECTIONS

Some studies suggest that a few standard interfaces should be the top priority for future development of vehicle location and navigation systems. These interfaces might include an application-layer protocol and a message set for traveler information services, a message standard for mayday systems, and an in-vehicle multiplex bus for ITS devices. A standard ITS data bus interface would allow the vehicle owner to add devices and the manufacturers and dealers to provide new features at much lower cost. Due to the typical design cycles of modern automobiles, it is difficult to install any recently developed electronic device in current-year models. Consumers must take what the vehicle manufacturers offer, or pay more later for an after-market retrofit. A standardized bus would allow easy system integration, so that existing vehicle devices could still be used. It would also allow plug and play with standard modules/connectors on the hardware side and standard application programming interfaces on the software side. It would transform the vehicle into a platform for third-party hardware and software products. It would be much easier to retrofit devices and systems or add new software features and applications. Remote access would be improved, too. It could permit remote diagnosis, upgrades, services, and feature additions. One such concept of an in-vehicle data bus is shown in Figure 12.1 [9].

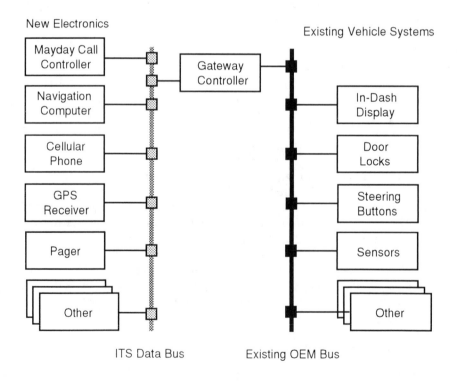

**Figure 12.1** Future in-vehicle ITS data bus concept.

Some believe that future vehicles should have more functions to enhance the safety of the driver and vehicle. According to this view, many safety and convenience technologies should be deployed along with the automatic highway system (AHS) [10–12]. These technologies include intelligent cruise control (space maintenance, speed control), collision avoidance, vision enhancement, and lane holding (Figure 12.2). Vehicle-to-vehicle communications would further strengthen the support of platooning and the basic roadway infrastructure. Vehicular travel on AHS would be automatically controlled without intervention from drivers or AHS operators. This would provide lanes with higher than normal capacity, in which the space or "headway" between vehicles would be reduced. The vehicles would revert to manual control as they exited the AHS. In addition, electronic (automatic) toll collection plazas and scales might be available along the highway to speed vehicles past toll booths and weigh stations.

Other studies predict that hand-held portable systems will be the navigation systems of the future. Assisted by various infrastructures, these systems would offer many of the same functions as in-vehicle systems [13]. A few early versions of portable systems have already reached the market, with limited functionality. Some use personal digital assistants (PDAs), notebook, or palmtop computers. Others use conventional paging services to provide traffic data. Different approaches exist. For instance, some available systems integrate a portable computer with a GPS receiver mounted on a PCMCIA card in conjunction with digital map and software packages. Some use static databases to provide travel information. In short, more advanced hand-held location and navigation systems will enter the consumer market.

Future automatic vehicle location (AVL) systems will also see many new changes. As wireless communications technologies rapidly advance, various new

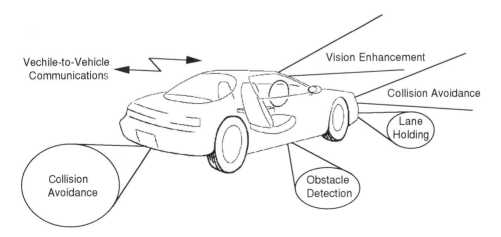

**Figure 12.2** Future concept vehicle.

wireless networks may become integrated with AVL systems. In-vehicle terminals may cost less, but still provide many new features. In addition to knowing where they were, law enforcement personnel would be able to access the Internet, WWW pages, or other databases in their vehicles to pull crime files and other information up on the screen in seconds. Many in-vehicle devices could be integrated into a single multifunction device or hand-held device. These new systems may lead to more efficient management of public transit, emergency fleet, hazardous material monitoring, access tracking, and many other public and private systems.

People around the world have similar visions for the future. In Europe, many anticipate the widespread use of the mobile GSM digital cellular phone both for communication and localization, the use of wireless personal travel assistants for intermodal travel, broadcasting of real-time traffic messages via FM radio (RDS-TMC), the development of intelligent cruise control, and a more powerful human-machine interface to display or digest the available information for navigation units. In Japan, people predict more integrated and miniaturized location and navigation systems with a better human-machine interface, much improved voice input devices, increased development and integration of data communications techniques to provide dynamic travel information, and more household media for vehicular use [14]. Navigation systems will become more and more integrated with existing entertainment equipment.

In the United States, mayday systems will continue their successful momentum, although autonomous vehicle navigation systems may eventually catch up as the unit price decreases. Rental cars will initially have the most widespread installed base. As the ITS infrastructure becomes more widely available, centralized navigation systems may gradually dominate the future market because of the benefits of knowing the real-time traffic information. Many autonomous systems might be ready for integration with centralized systems if good system architecture is used in conjunction with newly developed ITS standards. As location and navigation systems evolve, some in-vehicle systems may become more powerful, with additional features such as multimedia, e-mail, fax, and Internet surfing capability. Other systems, such as various mobile (hand-held) devices might become less vehicle dependent. For instance, GPS/cellular, GPS/CDPD, GPS/PDA, and various other technologies (terrestrial radio-based) might soon become available on a much larger scale for hand-held devices. Many of these future mobile devices may be motivated by the FCC's mandatory wireless E911 ruling (Section 10.4). They could be easily carried or plugged into the vehicle and might become either an independent or an integral part of the in-vehicle location and navigation unit while driving. These mobile devices might provide complementary data and exchange data with the in-vehicle navigation systems, home/office information systems, or public computer information terminals at major transport hubs or other information centers. All of these exciting developments will not be restricted to one particular region. In today's global economic environment, they will penetrate rapidly to different regions and eventually the entire world. As a result, there will be various systems with different services and upgrade capabilities from the very low end to the very high end for everyone to choose from.

Look to the future. The field of vehicle location and navigation is very promising. New and advanced technologies should be able to make transportation operate more safely and efficiently.

## References

[1] M. C. Paulk, C. V. Weber, B. Curtis, and M. B. Chrissis (Principal Contributors and Editors), *The Capability Maturity Model: Guidelines for Improving the Software Process*, Reading, MA: Addison-Wesley, 1995.

[2] F. J. Buckley, *Configuration Management: Hardware, Software, and Firmware*, New York: IEEE, 1993.

[3] S. B. Compton and G. R. Conner, in *Configuration Management for Software*, J. R. Callahan (Ed.), New York: Van Nostrand Reinhold, 1994.

[4] L. Sha, R. Rajkumar, and S. S. Sathaye, "Generalized Rate-Monotonic Scheduling Theory: A Framework for Developing Real-Time Systems," *Proc. IEEE*, Vol. 82, No. 1, Jan. 1994, pp. 68–82.

[5] R. Braek and O. Haugen, *Engineering Real Time Systems: An Object-Oriented Methodology Using SDL*, Englewood Cliffs, NJ: Prentice Hall, 1993.

[6] E. Gamma, R. Helm, R. Johnson, and J. Vlissides, *Design Patterns: Elements of Reusable Object-Oriented Software*, Reading, MA: Addison-Wesley, 1995.

[7] A. J. Riel, *Object-Oriented Design Heuristics,* Reading, MA: Addison-Wesley, 1996.

[8] M. Awad, J. Ziegler, and J. Kuusela, *Object-Oriented Technology for Real Time Systems: A Practical Approach Using OMT and Fusion*, Upper Saddle River, NJ: Prentice-Hall, 1996.

[9] A. M. Kirson and S. Scott, "A Reference Model for the Next Generation of In-Vehicle Electronics," *Abs. Third World Congress on Intelligent Transport Systems*, Paper #1064 (full paper in CD-ROM), Oct. 1996, p. 190.

[10] Joint Architecture Team, *ITS Architecture: Vision*, Federal Highway Administration, U.S. Department of Transportation, June 1996.

[11] NAHSC, *National Automated Highway System Consortium*, http:// web1.volpe.dot.gov/nahsc/, 1996.

[12] S. E. Underwood, K. Chen, and R. D. Ervin, "The Future of Intelligent Vehicle-Highway Systems: A Delphi Forecast and Sociotechnical Determinants," Transportation Research Board Paper No. 890804, U.S. Department of Transportation, 1989.

[13] R. L. French, "From Chinese Chariots to Smart Cars: 2,000 Years of Vehicular Navigation," *Navigation: J. Inst. Navigation*, Vol. 42, No. 1, Spring 1995, pp. 235–257.

[14] S. Azuma, K. Nishida, and S. Hori, "The Future of In-Vehicle Navigation Systems," *Proc. IEEE Vehicle Navigation and Information Systems Conference (VNIS '94)*, Aug./Sept. 1994, pp. 537–542.

# APPENDIX A

▼▼▼

# TRANSFORMATION BETWEEN CARTESIAN AND ELLIPSOIDAL COORDINATES

We first define the parameters of an ellipsoid as follows:

Semi-major axis:      $a$

Semi-minor axis:      $b$

Flattening:      $f = \dfrac{a - b}{a}$

Eccentricity:      $e = \dfrac{\sqrt{a^2 - b^2}}{a}, \text{ or } = \sqrt{1 - (1 - f)^2}$

Assume that an ellipsoid of revolution is placed at the origin of a Cartesian coordinate system and denote the Cartesian coordinates of a point in space by $X$, $Y$, $Z$ and the ellipsoidal coordinates of the same point by $\phi$, $\lambda$, $h$. The relationships between the Cartesian coordinates and the ellipsoidal coordinates will then be given by [1,2]

$$X = (N + h)\cos\phi\cos\lambda$$

$$Y = (N + h)\cos\phi\sin\lambda$$

$$Z = \left(\frac{b^2}{a^2}N + h\right)\sin\phi$$

where $N$ is the radius of curvature at the prime vertical which can be obtained by

$$N = \frac{a^2}{\sqrt{a^2\cos^2\phi + b^2\sin^2\phi}}, \text{ or } = \frac{a}{\sqrt{1 - e^2\sin^2\phi}}$$

The inverse transformations are derived as follows [3]:

$$\phi = \arctan\left[\frac{(2t + e^2)z}{(2t - e^2)\sqrt{x^2 + y^2}}\right]$$

$$\lambda = 2\arctan\left[\frac{\sqrt{x^2 + y^2} - x}{y}\right]$$

$$h = \text{sign}\left(t + \frac{e^2}{2} - 1\right)\sqrt{(x^2 + y^2)\left(1 - \frac{2}{2t + e^2}\right)^2 + z^2\left(1 - \frac{2(1 - e^2)}{2t - e^2}\right)^2}$$

where

$$t = \sqrt{\sqrt{\beta^2 - k} - \frac{(\beta + i)}{2}} - \text{sign}(m - n)\sqrt{\frac{(\beta - i)}{2}}$$

$$\beta = \frac{i}{3} - \sqrt[3]{q + D} - \sqrt[3]{q - D}$$

$$k = \frac{e^4}{4}\left(\frac{e^4}{4} - m - n\right)$$

$$i = -\frac{\left(\frac{e^4}{2} + m + n\right)}{2}$$

$$m = \frac{x^2 + y^2}{a^2}$$

$$n = \left[\frac{(1 - e^2)z}{b}\right]^2$$

$$q = \frac{(m + n - e^4)^3}{216} + \frac{mne^4}{4}$$

$$D = \sqrt{\left(2q - mn\frac{e^4}{4}\right)mn\frac{e^4}{4}}$$

Note that when $\sqrt{x^2 + y^2} = 0$

$$h = \text{sign}(z)z - b$$

$$\phi = \text{sign}(z)\frac{\pi}{2}$$

## References

[1] B. Hofmann-Wellenhof, H. Lichtenegger, and J. Collins, *Global Positioning System: Theory and Practice*, 3rd ed., Berlin: Springer-Verlag, 1994.

[2] A. Leick, *GPS Satellite Surveying*, 2nd ed., New York: John Wiley & Sons, 1995.

[3] J. Zhu, "Exact Conversion of Earth-Centered, Earth-Fixed Coordinates to Geodetic Coordinates," *J. Guidance, Control, Dynamics*, Vol. 16, No. 2, 1993, pp. 389–391.

# APPENDIX B
▼▼▼

# TRANSFORMATION BETWEEN UTM AND ELLIPSOIDAL COORDINATES

In the Universal Transverse Mercator (UTM) projection, based on the same ellipsoidal parameters defined in Appendix A, the transformation equations for transforming an ellipsoidal point $(\phi, \lambda)$ to a point $(y, x)$ on the UTM plane are given by the following series expansions [1,2]:

$$
\begin{aligned}
y = kS(\phi) + kN\Bigg[ &\frac{t}{2}\cos^2\phi(\lambda - \lambda_0)^2 + \frac{t}{24}\cos^4\phi(5 - t^2 + 9\eta^2 + 4\eta^4)(\lambda - \lambda_0)^4 \\
&+ \frac{t}{720}\cos^6\phi(61 - 58t^2 + t^4 + 270\eta^2 - 330t^2\eta^2)(\lambda - \lambda_0)^6 \\
&+ \frac{t}{40320}\cos^8\phi(1385 - 3111t^2 + 543t^4 - t^6)(\lambda - \lambda_0)^8 + \dots \Bigg]
\end{aligned}
$$

$$
\begin{aligned}
x = kN\Bigg[ &\cos\phi(\lambda - \lambda_0) + \frac{1}{6}N\cos^3\phi(1 - t^2 + \eta^2)(\lambda - \lambda_0)^3 \\
&+ \frac{1}{120}\cos^5\phi(5 - 18t^2 + t^4 + 14\eta^2 - 58t^2\eta^2)(\lambda - \lambda_0)^5 \\
&+ \frac{1}{5040}\cos^7\phi(61 - 479t^2 + 179t^4 - t^6)(\lambda - \lambda_0)^7 + \dots \Bigg]
\end{aligned}
$$

where

Arc length of meridian: $\quad\quad\quad\quad\quad$ $S(\phi)$

Radius of curvature in prime vertical: $\quad$ $N = \dfrac{a^2}{b\sqrt{1 + \eta^2}}$

Auxiliary quantity: $\quad\quad\quad\quad\quad\quad$ $\eta^2 = e'^2\cos^2\phi$

Second numerical eccentricity: $\quad\quad\quad$ $e'^2 = \dfrac{a^2 - b^2}{b^2}$

Auxiliary quantity: $\quad\quad\quad\quad\quad\quad$ $t = \tan\phi$

Longitude of the central meridian: $\quad\quad$ $\lambda_0$

UTM scale factor: $\quad\quad\quad\quad\quad\quad$ $k = 0.9996$

Because the pair of coordinates $(y, x)$ corresponds to $(\phi, \lambda)$, $y$ is given first followed by $x$. The arc length of the meridian $S(\phi)$ is the ellipsoidal distance from the equator to the point to be transformed and is given by the series expansion

$$S(\phi) = \alpha[\phi + \beta\sin2\phi + \gamma\sin4\phi + \delta\sin6\phi + \epsilon\sin8\phi + \ldots]$$

where

$$\alpha = \frac{a + b}{2}\left(1 + \frac{1}{4}n^2 + \frac{1}{64}n^4 + \ldots\right)$$

$$\beta = -\frac{3}{2}n + \frac{9}{13}n^3 - \frac{2}{32}n^5 + \ldots$$

$$\gamma = \frac{15}{16}n - \frac{15}{32}n^4 + \ldots$$

$$\delta = -\frac{35}{48}n^3 + \frac{105}{256}n^5 - \ldots$$

$$\epsilon = \frac{315}{512}n^4 + \ldots$$

$$n = \frac{a - b}{a + b}$$

As an example, the parameters of the WGS 84 reference ellipsoid (Table 2.1) are listed here:

$$\alpha = 6367449.1458\text{m}$$
$$\beta = -2.51882792 \times 10^{-3}$$
$$\gamma = 2.64354 \times 10^{-6}$$
$$\delta = -3.45 \times 10^{-9}$$
$$\epsilon = 5 \times 10^{-12}$$
$$n = 1.67922 \times 10^{-3}$$

The inverse UTM projection is the transformation of a point $y$, $x$ in the plane to a point $\phi$, $\lambda$ on the ellipsoid. The transformation equations are given by the following series expansions:

$$\phi = \phi_f - \frac{t_f}{2}(1 + \eta_f^2)\left(\frac{x}{kN_f}\right)^2$$

$$+ \frac{t_f}{24}(5 + 3t_f^2 + 6\eta_f^2 - 6t_f^2\eta_f^2 - 3\eta_f^4 - 9t_f^2\eta_f^4)\left(\frac{x}{kN_f}\right)^4$$

$$- \frac{t_f}{720}(61 + 90t_f^2 + 45t_f^4 + 107\eta_f^2 - 162t_f^2\eta_f^2 - 45t_f^4\eta_f^2)\left(\frac{x}{kN_f}\right)^6$$

$$+ \frac{t_f}{40320}(1385 + 3633t_f^2 + 4095t_f^4 + t_f^6)\left(\frac{x}{kN_f}\right)^8 + \cdots$$

$$\lambda = \lambda_0 + \frac{1}{\cos\phi_f}\left(\frac{x}{kN_f}\right) - \frac{1}{6\cos\phi_f}(1 + 2t_f^2 + \eta_f^2)\left(\frac{x}{kN_f}\right)^3$$

$$+ \frac{1}{120\cos\phi_f}(5 + 28t_f^2 + 24t_f^4 + 6\eta_f^2 + 8t_f^2\eta_f^2)\left(\frac{x}{kN_f}\right)^5$$

$$- \frac{1}{5401\cos\phi_f}(61 + 662t_f^2 + 1320t_f^4 + 720t_f^6)\left(\frac{x}{kN_f}\right)^7 + \cdots$$

where the terms with the subscript $f$ must be computed based on the footpoint latitude $\phi_f$. The series expansion of the footpoint latitude is given by

$$\phi_f = \frac{y}{k\alpha} + \overline{\beta}\sin 2\frac{y}{k\alpha} + \overline{\gamma}\sin 4\frac{y}{k\alpha} + \overline{\delta}\sin 6\frac{y}{k\alpha} + \overline{\epsilon}\sin 8\frac{y}{k\alpha} + \cdots$$

where

$$\bar{\alpha} = \frac{a+b}{2}\left(1 + \frac{1}{4}n^2 + \frac{1}{64}n^4 + \ldots\right)$$

$$\bar{\beta} = \frac{3}{2}n - \frac{27}{32}n^3 + \frac{269}{512}n^5 + \ldots$$

$$\bar{\gamma} = \frac{21}{16}n^2 - \frac{55}{32}n^4 + \ldots$$

$$\bar{\delta} = \frac{151}{96}n^3 - \frac{417}{128}n^5 + \ldots$$

$$\bar{\epsilon} = \frac{1097}{512}n^4 + \ldots$$

Note that the coefficient $\bar{\alpha}$ is identical to $\alpha$ in the previous transformation. As an example, the parameters for the WGS 84 reference ellipsoid are listed:

$$\bar{\alpha} = 6367449.1458 \text{ m}$$
$$\bar{\beta} = 2.51882658 \times 10^{-3}$$
$$\bar{\gamma} = 3.70095 \times 10^{-6}$$
$$\bar{\delta} = 7.45 \times 10^{-9}$$
$$\bar{\epsilon} = 17 \times 10^{-12}$$

Note that for location and navigation applications, the higher order terms indicated by ellipses can be ignored. Setting the UTM scale factor equal to one leads to the transformation equations for the transverse Mercator projection.

### References

[1] B. Hofmann-Wellenhof, H. Lichtenegger, and J. Collins, *Global Positioning System: Theory and Practice*, 3rd ed., Berlin: Springer-Verlag, 1994.

[2] A. Leick, *GPS Satellite Surveying*, 2nd ed., New York: John Wiley & Sons, 1995.

# APPENDIX C
▼▼▼

# POSITIONING SENSOR
# TECHNOLOGIES

To become familiar with the sensor technologies used to determine position, we have summarized the families and underlying principles of various types of linear and angular vehicular position sensors in Tables C.1 and C.2, respectively. Some sensors listed can be used for velocity measurement. Note that the boundaries of some sensor families are vague because they may have hybrid elements produced by mechanical, electrical, or other technologies. The tables attempt to illustrate the variety of technologies available. Because of rapid progress in sensor technologies, it is very difficult to cover them all in two tables. Many technologies can be used for either linear sensors or angular sensors. Certain technologies may not be suitable for vehicle positioning applications at the moment.

**Table C.1**
Linear Position Sensors

| Family | Sensor | Underlying Principle |
|--------|--------|----------------------|
| Mechanical | Microswitch | Simple on and off |
| | Reed switch | Activated when aligned with magnetic flux |
| Electromechanical | Eddy current | Currents induced in conductive object |
| | Hall effect | Voltage across semiconductor in magnetic field |
| | Linear variable differential transformer (LVDT) | Coupled transformers with moving core |
| | Magnetic transistor | Voltage across transistor in magnetic field |
| | Variable reluctance | Changes in reluctance within magnetic circuit |
| | Wiegand effect | Voltages induced by magnetic pulses |
| Electro-optical | Charge-coupled device (CCD) array, lateral effect photodiodes | Charge is proportional to incident light |
| | Laser | Time required to transmit and receive laser energy |
| | Light curtain | Through-beam array of source detector pairs |
| | Linear optical encoder | Source/detector pair interrupted by slits |
| | Interferometer | Optical phase |
| | Radar (microwave) | Time required to transmit and receive |
| Solid state | Magnetoresistive element | Resistance change in proportion to magnetic field strength |
| Acoustic | Ultrasonic | Time required to transmit and receive piezoelectric transducer pulses or Doppler effect |

**Table C.2**
Angular Position Sensors

| *Family* | *Sensor* | *Underlying Principle* |
|---|---|---|
| Mechanical | Gas rate gyroscope | Deflection of gas flow |
| | Rotor gyroscope | Conservation of angular momentum |
| | Vibration gyroscope | Coriolis acceleration |
| Electromechanical | Capacitor | Changing geometry or capacitance |
| | Fluxgate compass | Alter permeability in magnetic field to induce voltage |
| | Hall effect | Voltage across semiconductor in magnetic field |
| | Pin contact rotary encoder | Switch contact |
| | Rotary variable differential transformer (RVDT) | Coupled transformers with rotary core |
| | Synchro resolver | Three-phase transformer excited by ac rotor |
| Electro-optical | Fiber-optic gyroscope | Optical phase |
| | Optical encoder | Source/detector pair interrupted by slotted disk |
| Electronic | Electrolytic vial | Variable resistance between a liquid and three electrodes |
| Solid state | Magnetoresistive element | Resistance change proportional to magnetic field strength |

# APPENDIX D

▼▼▼

# LEAST SQUARES ALGORITHM

Given the following equation:

$$\mathbf{Y} = A\mathbf{X}$$

where $\mathbf{Y}$ is the known $n \times 1$ vector of measurements, $\mathbf{X}$ is the $m \times 1$ vector of unknowns, and $A$ is the $n \times m$ design matrix.

If $n > m$, the system has more equations than unknowns. The method of least squares can be used to obtain an optimal estimate for $\mathbf{X}$ using the redundant set of measurements.

Let us define the $n \times 1$ vector of residuals as follows:

$$\mathbf{r} = A\mathbf{X} - \mathbf{Y}$$

We now need to find the value for $\mathbf{X}$ that minimizes the sum squares of the residuals, that is,

$$f(\mathbf{X}) = (A\mathbf{X} - \mathbf{Y})^2 = (A\mathbf{X} - \mathbf{Y})^T(A\mathbf{X} - \mathbf{Y})$$

where $T$ indicates the transpose operation. Differentiating the preceding equation and setting the result to zero, we obtain

$$\frac{df(\mathbf{X})}{d\mathbf{X}} = 2A^T A\mathbf{X} - 2A\mathbf{Y} = 0$$

Provided $A^T A$ is nonsingular, we solve for $\mathbf{X}$ and obtain

$$\mathbf{X} = (A^T A)^{-1} A^T \mathbf{Y}$$

This is the least squares solution, where $(A^T A)^{-1}$ is the cofactor matrix we studied in Section 3.3. If $n = m$ and $A$ is nonsingular, the least squares solution reduces to the following unique solution:

$$\mathbf{X} = A^{-1}\mathbf{Y}$$

Note that the observations can be weighted based on the accuracy of each observation. The preceding solutions describe the equal-weight case. For more information, refer to [1,2].

### References

[1] G. F. Franklin, J. D. Powell, and M. L. Workman, *Digital Control of Dynamic Systems*, 2nd ed., Reading, MA: Addison-Wesley, 1990.
[2] M. Gopal, *Modern Control System Theory*, 2nd ed., New York: John Wiley & Sons, 1993.

# APPENDIX E

▼▼▼

# KALMAN FILTER ALGORITHM

The optimum combination of measured and estimated results at time $t_k$ is given by

$$\hat{\mathbf{x}}_k = \hat{\mathbf{x}}_k^- + \mathbf{K}_k(\mathbf{z}_k - \mathbf{H}_k\hat{\mathbf{x}}_k^-)$$

where $\hat{\mathbf{x}}_k^-$ denotes the a priori estimate. The "hat" denotes estimate and the superscript minus indicates that this is our best estimate prior to assimilating the measurement at time $t_k$.

The Kalman filter gain used in the equation just given can be written as

$$\mathbf{K}_k = \mathbf{P}_k^- \mathbf{H}_k^T(\mathbf{H}_k\mathbf{P}_k^-\mathbf{H}_k^T + \mathbf{R}_k)^{-1}$$

and the error covariance matrix associated with the optimal estimate is obtained from

$$\mathbf{P}_k = (\mathbf{I} - \mathbf{K}_k\mathbf{H}_k)\mathbf{P}_k^-$$

To recursively compute the Kalman filter gain for the next step, the predictions for the state estimate and covariance at the next step are given by

$$\hat{\mathbf{x}}_{k+1}^- = \phi_k \hat{\mathbf{x}}_k$$

$$\mathbf{P}_{k+1}^- = \phi_k \mathbf{P}_k \phi_k^T + \mathbf{Q}_k$$

In these equations the covariance matrices $\mathbf{Q}_k$ and $\mathbf{R}_k$ are obtained from the $\mathbf{w}_k$ (process noise) and $\mathbf{v}_k$ (measurement noise) vectors as follows:

$$E\left[\mathbf{w}_k \mathbf{w}_i^T\right] = \begin{cases} \mathbf{Q}_k, & i = k \\ 0, & i \neq k \end{cases}$$

$$E\left[\mathbf{v}_k \mathbf{v}_i^T\right] = \begin{cases} \mathbf{R}_k, & i = k \\ 0, & i \neq k \end{cases}$$

$$E\left[\mathbf{w}_k \mathbf{v}_i^T\right] = 0, \text{ for all } k \text{ and } i$$

and, for the error covariance matrix $\mathbf{P}_k$,

$$\mathbf{P}_k^- = E\left[(\mathbf{x}_k - \hat{\mathbf{x}}_k^-)(\mathbf{x}_k - \hat{\mathbf{x}}_k^-)^T\right]$$

To derive the preceding estimation equations, we assume that the random process to be estimated can be modeled in the form

$$\mathbf{x}_{k+1} = \phi_k \mathbf{x}_k + \mathbf{w}_k$$

and the measurement (observation) of the process can be modeled as

$$\mathbf{z}_k = \mathbf{H}_k \mathbf{x}_k + \mathbf{v}_k$$

Detailed discussions of this algorithm can be found in [1–3].

### References

[1] R. G. Brown and P. Y. C. Huang, *Introduction to Random Signal Analysis and Applied Kalman Filtering*, New York: Wiley, 1992.
[2] A. Gelb (Ed.), *Applied Optimal Estimation*, Cambridge, MA: MIT Press, 1974.
[3] P. S. Maybeck, *Stochastic Models, Estimation and Control*, New York: Academic Press, 1979.

# APPENDIX F
▼▼▼

# FUZZY LOGIC BACKGROUND

Fuzzy set theory, introduced by Zadeh [1] in 1965, has been shown to be an effective way to deal with qualitatively oriented terms and concepts. Fuzzy logic provides a simple way to draw definite conclusions from vague, imprecise information. In an intuitive sense, it resembles human decision making. For example, consider the following declarative assertion concerning a vehicle traveling on a local residential road:

"The speed of this vehicle is high."

If the speed is 100 km/h, this statement is completely true. However, if 100 km/h is high, how about 99, 98, ... 90, ... 80, ... 70? At what degree does the statement suddenly become false? Since the speed can change gradually during a trip, it is hard to find a fixed value such that larger than that fixed value is "high" and less than that value is "not high." To resolve this fuzziness, we can assign a grade, or membership degree, to a speed. A speed is high only to a certain degree of truth, called a grade, or membership degree. By correlating various speeds to grades, we turn the preceding statement into a fuzzy membership function. Therefore, people often say that fuzzy logic is a mathematical formalism intended to handle these kinds of imprecise statements. It is different from traditional mathematical formalisms because fuzzy logic allows us to model the vagueness in our mental decision-making process. Consequently, fuzzy logic is intuitive and easier to work with than traditional logic for many applications. Traditional logic would have a difficult time processing the given statement. In the design of fuzzy inference systems, there are three basic concepts: membership functions, fuzzy logic operations, and fuzzy rules.

For membership functions, we can denote the grade of a speed in high by assigning it a numerical "truth value." Truth values may range anywhere between 0 and 1 and are typically considered to have meaning as suggested in Table F.1, although it may not always have meaning. In this manner we can arbitrarily assign various grades to "high" as a function of the speed (Table F.2).

The membership functions given in the tables are illustrated in Figure F.1. The curve in the figure illustrates how the membership function of a fuzzy concept is dependent on a measured value, for example, speed. When the relevant declarative assertions are combined into a single chart, more complicated curves are obtained. These curves are called fuzzy sets. The fuzzy concept "high" is called a label. For a speed, such as 70 km/h, 0.7 is its grade in high. In other words, if the speed of this

**Table F.1**
Linguistic Meanings of Fuzzy Logic Values

| Truth Value | Meaning |
| --- | --- |
| 0.0 | Completely false |
| 0.2 | Mostly false |
| 0.4 | Somewhat false |
| 0.6 | Somewhat true |
| 0.8 | Mostly true |
| 1.0 | Completely true |

**Table F.2**
Membership Function

| Speed (km/h) | Grade in High |
| --- | --- |
| 0 | 0.0 |
| 10 | 0.1 |
| 20 | 0.2 |
| 30 | 0.3 |
| 40 | 0.4 |
| 50 | 0.5 |
| 60 | 0.6 |
| 70 | 0.7 |
| 80 | 0.8 |
| 90 | 0.9 |
| 100 | 1.0 |

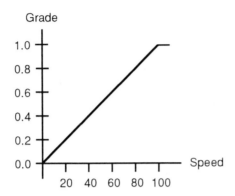

**Figure F.1** Membership function chart.

vehicle is 70 km/h, the truth value of the assertion "The speed of this vehicle is high" is 0.7.

For fuzzy logic operations, consider our spoken language. In addition to the simple assertion statement presented earlier, logical operators are often used in spoken language to combine simple assertions into more complex ones. Examples of such operators are "AND" and "OR." Assertions using these operators are called compound assertions. Similar operators can be used in fuzzy logic to help evaluate the truth values of these assertions. These operations are called fuzzy logic operations. For example, let us define another simple assertion:

"The distance to the next maneuver is short."

We can now create a compound assertion:

"The speed of this vehicle is high AND the distance to the next maneuver is short."

Once a truth value is assigned to each simple assertion, fuzzy logic operations can precisely evaluate the truth values of compound assertions. Even though the compound assertion is vague, the evaluation results are crisp values. In other words, nothing is fuzzy in the actual fuzzy inference process.

As examples, let us examine how to evaluate truth values for statements containing the two simple operators, AND or OR. Mathematical functions are used by fuzzy logic to evaluate two arbitrary statements A and B. There statements can be either simple or compound. The mathematical functions are listed in the right-hand column of Table F.3. Each of them is associated with its corresponding fuzzy logic operator in the left-hand column.

From the definitions given in Table F.3, we can easily evaluate the truth value of any statement. Let the truth value of A be 0.1 and that of B be 0.8. We obtain

**Table F.3**
Fuzzy Logic Operations

| Compound Operator on Statements A, B | Mathematical Function to Evaluate Truth |
|---|---|
| A AND B | minimum (min) |
| A OR B | maximum (max) |

Assertion 1: A AND B = min(0.1, 0.8) = 0.1

Assertion 2: A OR B  = max(0.1, 0.8) = 0.8

The crisp truth value of Assertion 1 is 0.1, whereas Assertion 2 has a value of 0.8.

We now know that a traditional logic such as Boolean logic is a special case of fuzzy logic. If you doubt this statement, see Table F.4, which lists various evaluation results for two compound assertions. Both results obtained by applying fuzzy logic operators and Boolean logic operators are derived. Note that both fuzzy logic and Boolean logic produce the same results when the truth values of the simple assertions are limited to 1 and 0. However, for values other than 1 or 0, fuzzy logic produces meaningful results while Boolean logic has no such operations defined.

For fuzzy rules, consider the following example. Suppose that we want to determine when to announce a maneuvering instruction via a speaker to a driver during route guidance. One of rules could be

IF the speed of this vehicle is high AND the distance to the next maneuver is short
THEN announce the next maneuver instruction.

**Table F.4**
Comparison of Boolean Logic and Fuzzy Logic Operations

| Logic Operator | | Fuzzy Logic | | Boolean Logic | |
|---|---|---|---|---|---|
| A | B | A AND B | A OR B | A AND B | A OR B |
| 0 | 0 | 0 | 0 | 0 | 0 |
| 0 | 1 | 0 | 1 | 0 | 1 |
| 1 | 0 | 0 | 1 | 0 | 1 |
| 1 | 1 | 1 | 1 | 1 | 1 |
| 0 | 0.6 | 0 | 0.6 | ? | ? |
| 0.5 | 0.5 | 0.5 | 0.5 | ? | ? |
| 0.8 | 0.6 | 0.6 | 0.8 | ? | ? |
| 1 | 0.6 | 0.6 | 1 | ? | ? |

In other words, a rule has the following format:

IF <condition> THEN <consequence>.

The condition part of a rule could be either a simple or compound statement and the consequence part is typically a simple statement. The input to a fuzzy-logic-based system is very similar to human language. The more complex a system is, the more rules it will generally have. Given the input grades in the condition part, each rule is evaluated to obtain an output grade in the consequence part.

The final result of this inference process depends on the combinational effect of many fuzzy rules in the process. Because a single number is desired as output, we need a defuzzifying operation. In other words, a defuzzification operator is used to compute one final crisp output value, which is to make a decision for the system to follow. We can view this final step as interpreting the evaluated grades as a measure of confidence in the corresponding conclusion. The most popular method used for this operation is to compute the centroid (weighted average) of the output from the fuzzy rules. By considering the grades as weights attached to the assertions, the final decision $z$ is obtained by determining the weighted average of the support values associated with each output label:

$$z = \frac{\sum_{i=1}^{n} x_i y_i}{\sum_{i=1}^{n} y_i}$$

where $x$ is the position of the fuzzy output from the evaluation, $y$ is the value (weight) of the corresponding fuzzy output, and $n$ is the number of fuzzy outputs. In general, the outputs of fuzzy rule evaluation results are discrete values. For convenience, a curve is used in Figure F.2 to explain the basic concept.

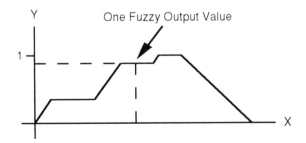

**Figure F.2** Example of a fuzzy output curve.

To summarize, the fuzzy logic inference process consists of the following three steps:

1. Fuzzification;
2. Rule evaluation;
3. Defuzzification.

As discussed before, fuzzification converts the input values into grades (fuzzy values) using input labels. This is often called relating crisp values to fuzzy sets by the membership functions of input labels. The rule evaluation step computes the grade of each output label, that is, the fuzzy output truth values are computed in this step. In general, this step involves rather complicated operations. It uses a inference engine with the membership functions of fuzzy sets to perform fuzzy reasoning on the rules. Defuzzification involves interpreting these grades as a measure of the confidence in the corresponding conclusions. It makes the final decision for the whole inference process. In other words, it converts the outcome of the fuzzy reasoning process into a crisp value that can be used to control the system or it can merely be presented to the system.

Different fuzzy logic operators are used for processing and interpreting rules and different defuzzification operators are used for deriving the final decision. In addition to AND (minimum) and OR (maximum), other examples of logic operators include product, probability sum, bounded intersection, bounded union, etc. Examples of defuzzification operators are centroid, leftmost maximizer, rightmost maximizer, average maximizer, etc. Details on the operators and fuzzy inference methods that are available can be found in [2,3].

## References

[1] L. A. Zadeh, "Fuzzy Sets," *Information and Control*, Vol. 8, No. 3, 1965, pp. 338–353.

[2] G. J. Klir and B. Yuan, *Fuzzy Sets and Fuzzy Logic: Theory and Applications*, Upper Saddle River, NJ: Prentice Hall, 1995.

[3] T. J. Ross, *Fuzzy Logic with Engineering Applications*, New York: McGraw-Hill, 1995.

▼▼▼

# ABOUT THE AUTHOR

Yilin Zhao received a B.E. in electrical engineering in 1982 from Dalian University of Technology, Dalian, People's Republic of China, and an M.S.E. in 1986 and Ph.D. in 1992, both from the Department of Electrical Engineering and Computer Science, University of Michigan, Ann Arbor.

In his early career, Dr. Zhao held managerial positions for work units of more than 100 people. He was an instructor from 1982 to 1984 and in 1995 was appointed as Adjunct Professor in the Department of Computer Science and Engineering, Dalian University of Technology. From 1987 to 1991, he was a teaching assistant and research assistant at the University of Michigan, where he studied and conducted research on autonomous vehicle control and navigation as well as intelligent transportation systems (ITS).

Dr. Zhao joined Motorola, Inc., in 1992. As a senior research and development engineer, he researches and develops vehicle location and navigation systems and automatic place-and-route systems for designing integrated circuit chips. His current interests include ITS, real-time computer systems, automatic placement and routing for large-scale integrated circuit design, and autonomous vehicle control and navigation. Dr. Zhao is a member of IEEE, the Mobile Robots Technical Committee of the IEEE Robotics and Automation Society, and the IEEE Technical Activities Board Intelligent Transportation Systems Committee.

▼▼▼

# INDEX

# The Artech House Mobile Communications Series

*John Walker, Series Editor*

*Wireless Communications in Developing Countries: Cellular and Satellite Systems,* Rachael E. Schwartz

*Wireless Communications for Intelligent Transportation Systems,* Scott D. Elliott, Daniel J. Dailey

*Wireless Data Networking,* Nathan J. Muller

*Wireless: The Revolution in Personal Telecommunications,* Ira Brodsky

For further information on these and other Artech House titles, contact:

Artech House
685 Canton Street
Norwood, MA 02062
617-769-9750
Fax: 617-769-6334
Telex: 951-659
email: artech@artech-house.com

Artech House
Portland House, Stag Place
London SW1E 5XA England
+44 (0) 171-973-8077
Fax: +44 (0) 171-630-0166
Telex: 951-659
email: artech-uk@artech-house.com

WWW: http://www.artech-house.com